Highlights cultural difference over neoliberalism

NEOLIBERALISM AND TECHNOSCIENCE

Theory, Technology and Society

Series Editor: Ross Abbinnett, University of Birmingham, UK

Theory, Technology and Society presents the latest work in social, cultural and political theory, which considers the impact of new technologies on social, economic and political relationships. Central to the series are the elucidation of new theories of the humanity-technology relationship, the ethical implications of techno-scientific innovation, and the identification of unforeseen effects which are emerging from the techno-scientific organization of society.

With particular interest in questions of gender relations, the body, virtuality, penality, work, aesthetics, urban space, surveillance, governance and the environment, the series encourages work that seeks to determine the nature of the social consequences that have followed the deployment of new technologies, investigate the increasingly complex relationship between 'the human' and 'the technological', or addresses the ethical and political questions arising from the constant transformation and manipulation of humanity.

Other titles in this series

Bio-Objects
Life in the 21st Century
Edited by Niki Vermeulen, Sakari Tamminen and Andrew Webster
ISBN 978 1 4094 1178 9

Decentering Biotechnology
Assemblages Built and Assemblages Masked
Michael S. Carolan
ISBN 978 1 4094 1005 8

The Genome Incorporated
Constructing Biodigital Identity
Kate O'Riordan
ISBN 978 0 7546 7851 9

Technology and Medical Practice
Blood, Guts and Machines
Edited by Ericka Johnson and Boel Berner
ISBN 978 0 7546 7836 6

Neoliberalism and Technoscience

Critical Assessments

Edited by

LUIGI PELLIZZONI
University of Trieste, Italy

MARJA YLÖNEN
University of Jyväskylä, Finland

ASHGATE

Published by
Ashgate Publishing Limited
Wey Court East
Union Road
Farnham
Surrey, GU9 7PT
England

Ashgate Publishing Company
Suite 420
101 Cherry Street
Burlington
VT 05401-4405
USA

www.ashgate.com

British Library Cataloguing in Publication Data
Neoliberalism and technoscience : critical assessments. — (Theory, technology and society)
1. Technology—Political aspects. 2. Science—Political aspects.
3. Technology—Social aspects. 4. Science—Social aspects. 5. Neoliberalism.
I. Series II. Pellizzoni, Luigi. III. Ylönen, Marja.
306.4'5–dc23

Library of Congress Cataloging-in-Publication Data
Pellizzoni, Luigi.
Neoliberalism and technoscience : critical assessments / by Luigi Pellizzoni and Marja Ylönen.
p. cm. — (Theory, technology, and society)
Includes bibliographical references and index.
ISBN 978-1-4094-3532-7 (hardback) — ISBN 978-1-4094-3533-4 (ebook)
1. Technological innovations—Economic aspects. 2. Technological innovations—Environmental aspects. 3. Neoliberalism. I. Ylönen, Marja. II. Title.
HC79.T4P45 2012
303.48'3–dc23

2012007262

ISBN 9781409435327 (hbk)
ISBN 9781409435334 (ebk)

MIX
Paper from responsible sources
FSC® C018575
FSC www.fsc.org

Printed and bound in Great Britain by the MPG Books Group, UK.

Contents

PART 3: Neoliberalism, technoscience and the environment

Notes on contributors

Simone Arnaldi is Research Fellow at the Centre for Environmental Law Decisions and Corporate Ethical Certification of the University of Padova, Italy. He studies the way images of future technologies affect innovation discourse and practice, and, in general, the relationships between science, technology, and society.

Arthur Arruda Leal Ferreira is a post-doctoral researcher in the History of Psychology at the UNED (Spain) and Professor of the History of Psychology at the Institute of Psychology at the Federal University of Rio de Janeiro (UFRJ). He has recently edited the following books: *A pluralidade do campo psicológico, História da Psicologia: Rumos e Percursos*, *Teoria Ator-Rede e a Psicologia* and *Pragmatismo e questões contemporâneas*. He also contributed to the books: *Foucault Hoje*, *Foucault e a Psicologia*, *Da metafísica moderna ao pragmatismo* and *Psicologia e Instituições no Brasil.*

Imre Bárd is Research Officer at the London School of Economics and Political Science, working on various projects related to the social and ethical aspects of neuroenhancement. He has a degree in Philosophy from the University of Vienna and an MSc in Biomedicine, Bioscience and Society from the LSE.

Kean Birch is Assistant Professor in Department of Social Science (Business and Society Program) at York University, Toronto. His current research interests range from the emerging bioeconomy to varieties of neoliberal restructuring. With Vlad Mykhnenko he co-edited *The Rise and Fall of Neoliberalism* (2010) published by Zed Books.

Anders Blok is Associate Professor at the Department of Sociology, Copenhagen University, Denmark. His current research is on the knowledge and governance of urban climate change planning, based on case studies of large-scale cities in three different regions of the world. Main fields of expertise include environmental sociology and science & technology studies (STS), especially around transnational issues of biodiversity and climate change politics. He has published widely on these topics in leading academic journals, including *Science, Technology & Human Values* and *Economy and Society*. Apart from Copenhagen, Anders Blok did post-graduate sociology training at the universities of Lancaster (UK) and Harvard (US), and he has extensive research experience in Japan and India.

Rodrigo Costa Nascimento holds a degree in Psychology from the Institute of Psychology at the Federal University of Rio de Janeiro (UFRJ), and a Master's degree in Collective Health from the Institute of Collective Health Studies of the same university.

Alessandro Gandini is Ph.D. Candidate in Sociology at University of Milan. He is currently working on his Ph.D. dissertation on reputation economy in creative labour markets, focusing on freelance work. His main research interests also include digital media, social network analysis and social innovation.

David J. Hess is Professor of Sociology at Vanderbilt University. He has recently written a series of books that describe the pressure points in the political system where political opportunities can be opened for a more rapid green transition in the United States. The first book, *Alternative Pathways in Science and Industry*, focused on social movements and associated research programs; the second book, *Localist Movements in a Global Economy*, studied the potential for local economic ownership to enhance sustainability and justice goals; and the third book, *Good Green Jobs in a Time of Crisis*, studied developmentalism and green-transition coalitions.

Les Levidow is Senior Research Fellow at the Open University, UK, where he has been studying agri-food-environmental issues since the late 1980s. His research topics have included the following: sustainable development, agri-food-energy innovation, agricultural research priorities, governance, European integration, regulatory expertise and the precautionary principle. He is co-author of two books: *Governing the Transatlantic Conflict over Agricultural Biotechnology: Contending Coalitions, Trade Liberalisation and Standard Setting* (Routledge, 2006); and *GM Food on Trial: Testing European Democracy* (Routledge, 2010). Other publications are listed at http://dpp.open.ac.uk/people/levidow.htm. He is also Editor of the journal *Science as Culture*.

Karina Lopes Padilha holds a degree in Psychology from the Institute of Psychology at the Federal University of Rio de Janeiro (UFRJ), is a specialist in Collective Health at the Institute of Collective Health Studies of the same University, and a master's student in Collective Health at the Institute of Social Medicine, State University of Rio de Janeiro (UERJ).

Theo Papaioannou is Senior Lecturer in Innovation and Politics of Development at the ESRC Centre for Social and Economic Research on Innovation in Genomics (INNOGEN) and the Development Policy and Practice Group (DPP) at the Open University, UK. He is currently Director of Ethics and Politics of Life Sciences Innovation at INNOGEN and Leader of the Governance of Innovation and Development Research Stream at IKD. He has researched and published extensively in the areas of innovation, politics and development. He co-edited *The*

Limits to Governance: The Challenge of Policy-Making for the New Life Sciences, Aldershot: Ashgate (2009).

Luigi Pellizzoni is Associate Professor in Environmental and Political Sociology at the University of Trieste, Italy. His research interests connect risk and uncertainty, environment and technoscience, social conflicts and new forms of governance. On these topics he has has published, among others, in the *British Journal of Sociology*, *European Journal of Social Theory*, *Theory Culture and Society*, *Environmental Politics*, *Environmental Values*, *Global Environmental Change*. Recent books include the co-authorship of *Il rischio ambientale* (Il Mulino, 2001) and *Sociologia dell'Ambiente* (Il Mulino, 2008), and the editorship of *La Deliberazione Pubblica* (Meltemi, 2005) and *Conflitti Ambientali. Esperti, Politica e Istituzioni nelle Controversie Ecologiche* (Il Mulino, 2011).

Laurence Reynolds is Research Fellow at the ESRC Centre for Economic and Social Aspects of Genomics (Cesagen) and the Sociology Department at Lancaster University. His work addresses the relationships between technology, the environmental crisis and capitalism.

Míriam Starosky holds a degree in is Psychology from the Institute of Psychology, Federal University of Rio de Janeiro (UFRJ) and is a Master's student in Sociology at the Institute of Social and Political Studies at the State University of Rio de Janeiro (UERJ).

Bronislaw Szerszynski is Senior Lecturer in the Department of Sociology at Lancaster University, UK, where he also works in the Centre for the Study of Environmental Change (CSEC) and the ESRC Centre for the Economic and Social Aspects of Genomics (Cesagen). His current research topics include geoengineering, local food networks and (with Laurence Reynolds) a critical appraisal of the possibility of a transition to a green techno-economic paradigm. He is author of *Nature, Technology and the Sacred* (Blackwell, 2005), and co-editor of *Risk, Environment and Modernity* (Sage/TCS, 1996), *Re-Ordering Nature: Theology, Society and the New Genetics* (T&T Clark, 2003) and *Nature Performed: Environment, Culture and Performance* (Blackwell, 2003).

Marja Ylönen is post-doctoral researcher at the University of Jyväskylä, Finland. She has taught environmental sociology, qualitative methods and sociology of deviance. She has co-authored an environmental sociology textbook and her recent dissertation deals with an ideology of social control of water pollution and pollution crimes in Finland from the 1960s until 2000. Her current research topic concerns regulatory challenges to nuclear safety after the Fukushima disaster. Previous work with Luigi Pellizzoni includes an article published in *Global Environmental Politics*.

Introduction

Luigi Pellizzoni and Marja Ylönen

Exploring the connections between neoliberalism and technoscience is more challenging than it may seem at first sight. The temporal overlap between the spread of neoliberalization programmes and policies, the institutional and cultural reorganization of science, and the novel technoscientific wave, which focuses on biotechnologies, ICTs, neurosciences and nanotechnologies, would seem to make such inquiry reasonably straightforward. Yet venturing into the terrain of neoliberalism is a risky endeavour, even at a conceptual level. While the notion of technoscience has long been accepted as a synthetic descriptor of the increasingly tight coupling of science and technology, the term neoliberalism has a much more problematic status. Its normative implications, for a start, are stronger, or more obvious. Moreover, as Bourdieu and Wacquant (2001) remark, neoliberalism has become a sort of 'planetary vulgate'. This, for some, affects the analytical value of the notion, with too many or too vague, if not contradictory, understandings and uses (see e.g. Barnett 2005). The concept, in other words, would be 'too big or bloated to capture the diversity of projects labelled as neoliberal' (Heynen et al. 2007a: 4). The sheer bulk of the literature devoted to the topic – in a recent survey Peck, Theodore and Brenner (2009) found about 2,500 English-language articles in the social sciences that cite 'neoliberalism' as a keyword, 86% of which were published after 1998 – hardly seems to constitute an argument in favour of further lucubration.

Other scholars, however, contend that the notion of neoliberalism can actually perform a valuable function as a reference point for critical inquiry (Ward and England 2007, Heynen et al. 2007a). We agree, for at least two reasons. One is that the very plurality of meanings and readings of neoliberalism is positive to the extent that it draws attention to the dynamic, complex nature of the issue. Many actually prefer to talk of 'neoliberalization', rather than neoliberalism, precisely in order to stress the processual, manifold, ever-changing character of the phenomenon. The other reason is that, compared with closely-related concepts like capitalism, globalization, knowledge economy or knowledge society, use of the notion of neoliberalism means stressing the relevance of the interplay of technoscience with *political* rationalities in the first instance. To talk of neoliberalism, in other words, means choosing a specific perspective on the current state of affairs in technoscience, which can be rejected, but not dismissed as irrelevant. If anything, its significance is confirmed by the responses of governments to the recent economic-financial crises, which reiterate approaches that, according to most commentators, are at the very origin of the crises themselves. In the words of Colin Crouch, 'neoliberalism is emerging from the financial collapse more politically powerful than ever' (Crouch 2011: viii).

It is not an overstatement, therefore, to regard neoliberalism as a far-reaching, all-encompassing affair – a process of institutional, cultural and social change of 'a scale not seen since the immediate aftermath of the Second World War' (Campbell and Pedersen 2001: 1). It would be quite surprising, then, to find no or merely casual connections between neoliberalization processes and technoscience. Yet despite widespread acknowledgment of the major role innovation plays in post-Fordist, globalized economy, a relatively small amount of literature specifically addresses the neoliberalism–technoscience relationship.

The story of neoliberalism has been told many times, and one can find excellent assessments in monographs such as those by Harvey (2005), Peck (2010), Steger and Roy (2010) and Crouch (2011), or edited volumes such as those by Saad-Filho and Johnston (2005) England and Ward (2007), Heynen et al. (2007c), and Birch and Mykhnenko (2010a). Several aspects of this story are discussed in the chapters of this book. In this introduction we provide an overview of some of the major points of debate, looking at the way neoliberalism and its connections with technoscience have been thematized, and outlining the basic aims, structure and contents of the book. In the concluding chapter we shall offer some reflections on the picture which emerges from this collection of essays.

Neoliberalism and neoliberalization processes

Neoliberalism as an ideology dates back to the late 1930s, when certain liberal intellectuals endeavoured to rework classic liberalism in response to totalitarianism and Keynesian state planning. The prominent figures were Ludwig von Mises and Friedrich von Hayek, both from the Austrian school of economics. After the Second World War, neoliberalism expanded internationally through intellectual-political networks such as the Mont Pelerin Society and the Davos World Economic Forum (Mirowski and Plehwe 2009). The University of Chicago, where a school of neoliberal economists (the 'Chicago Boys') flourished around the figure of Milton Friedman, became a major centre of development. Over the years, institutions such as the International Monetary Fund (IMF), the World Bank, the World Trade Organization (WTO) and the Organization for Economic Cooperation and Development (OECD) played a major role in triggering the translation of neoliberal ideas into actual programmes and policies. Pivotal moments in this transition have been the fall of the Bretton Woods regime of fixed exchange rates, in 1971, the catalyst for which was US expenditure on the Vietnam War; the New York City fiscal crisis of 1975, when the employees' pension funds were used for the first time to buy corporate bonds (issued by the Municipal Assistance Corporation); the Pinochet coup in Chile in 1973, with the subsequent anti-socialist reforms which were worked out by some of the 'Chicago Boys'; the advent of the Reagan and

Thatcher administrations at the turn of the 1980s; the meltdown of the Soviet bloc during the following decade.

Most analyses concur in depicting neoliberalization processes as a response to crises (Jessop 2002) – initially to the crisis of the 'embedded liberalism' which focused on the Keynesian state and the Fordist economy, and subsequently to the crises provoked by neoliberal reforms themselves. The stagflation crisis of the 1970s, with sustained public deficit and monetary inflation, declining returns on capital investments, and the expansion of salaries to the detriment of profits, 'provided the political opportunity to push for a new economic project founded on neoliberal assumptions about economic efficiency, reduced state intervention and free markets' (Birch and Mykhnenko 2010b: 4). Neoliberalization has never been an organic, consistent process, however. It is, rather, a complex, contested, contradictory assemblage of policies, practices and discourses (Ong 2006, Birch and Mykhnenko 2009). On the one hand, it has been seen that there are diverse ways of reacting – promoting, adjusting or resisting – to the 'global neoliberal turn' (Jessop 2002). On the other, there are different phases or waves of neoliberalization (Brenner, Peck and Theodore 2010). A common distinction in this respect – based on the Polanyian idea of 'double movement' (Polanyi 1944): that is, the historical oscillation between the spread of market capitalism and counteractions aimed at limiting its socially disruptive effects – is between the 'roll-back' phase of the 1980s and early 1990s, which focused on deregulation and the dismantling of the welfare state, and a subsequent 'roll-out' phase concerned with establishing market-guided state regulation and promoting non-market metrics, social capital, local governance and public–private partnerships (Peck and Tickell 2002). The latter period corresponds with Tony Blair's 'third way' in the UK and Bill Clinton's 'market globalism' in the US – in practice, an attempt to incorporate parts of a socially progressive agenda while maintaining a basically neoliberal framework (Steger and Roy 2010: 50 ff.).

The varieties of 'actually existing neoliberalism' (Brenner and Theodore 2002) have triggered a good deal of criticism of 'ahistorical and ageographical invocations of a poorly defined abstraction' (Heynen et al. 2007a: 4), with special reference to an alleged 'hegemonic', all-encompassing class project (Plehwe, Walpen and Neunhöffer 2005). This criticism does not, however, generally contest the evidence of world-wide institutional and regulatory thrusts and counter-thrusts (Jessop 2002, Arrighi and Silver 2001). It makes sense, therefore, to search for a basic unity behind the complexity and contradiction, to account for how neoliberalism 'has gone truly global, reaching every corner of the world' (Birch and Mykhnenko 2010b: 8); to account for how and why it is that, 'given the interdependencies and interrelatedness of agency, contingency, and complexity, [...] people willingly choose the same items from the same tiny menu over and over' (Heynen et al. 2007b: 289). Moreover, if one considers the growing concentration of wealth which has taken place across the world in recent decades (OECD 2008), it is hard to dismiss the argument that the current handling of social affairs is more than the patchwork of responses to 'objective' challenges of economic globalization and technological innovation which the literature building on the concept of 'governance' often suggests. Rather, it shows elements of 'an intensely political project, one in which economic elites more or less intentionally seek to increase their wealth and income, but also their political and economic freedom and flexibility' (Heynen et al. 2007a: 5). Interpreting neoliberalism as a project aimed at the restoration of class privilege and power which had been partially lost in the period of embedded liberalism, however, means maintaining neither that this project has been pursued consistently, nor that the social groups who benefit from it have remained the same – innovation, especially in the ICT and biotechnology fields, has been a major trigger for change in industrial and financial elites in the past three decades.

For Harvey, the essence of neoliberalism is the assumption that 'human well-being can best be advanced by liberating individual entrepreneurial freedoms and skills within an institutional framework characterized by strong private property rights, free markets, and free trade' (Harvey 2005: 2). At its most basic level, neoliberalism shows a preference for markets as policy instruments, for trade liberalization over protectionism, and for self-enterprising and responsibility (Moore et al. 2011). In concrete, neoliberal policies are often more devoted to the protection of corporate power than of free markets as such (Crouch 2011). The policy package of neoliberalization is famously synthesized in the 'Washington Consensus' list of reforms proposed by the economist John Williamson (1993): fiscal discipline (no public budget deficit), tax cuts, financial liberalization, free-floating exchange rates, trade liberalization, the promotion of foreign investments, the reduction of public expenditure, privatization, the deregulation of labour and product markets, and the strengthening of property rights. To expand on this list, one may add a general marketization of society through public–private partnerships and other forms of commodification; public expenditure focused on 'workfare' and a supply-side economy (support for development, innovation and competitiveness, according to a Ricardian emphasis on competitive specialization at the level of countries, regions, enterprises and individuals); re-regulation in terms of offloading and decentralization of responsibilities, with re-scaling of government up and down from nation-states; an emphasis on contracts and transaction-intensive markets; reciprocal support of neoliberal and neoconservative agendas, with convergence on 'family values', tough law enforcement, and a strong military (Harvey 2005, Brown 2006, Heynen et al. 2007a, Ward and England 2007, Bumpus and Liverman 2008, Birch and Mykhnenko 2010b, Steger and Roy 2010).

Neoliberalization can also be seen as a major phase of 'accumulation by dispossession' (Harvey 2003), by which elites respond to problems of capital over-accumulation by searching for new, or less expensive, areas of investment, beginning with land and raw materials, to which renewed processes of enclosure, commodification and marketization are applied. Neoliberalism, as a consequence, prominently affects biophysical nature with the crucial contribution of scientific knowledge and technological means. In this sense, neoliberalism is 'also an *environmental* project, and *necessarily* so. [It] tends not only to generate serious environmental consequences but [...] is significantly constituted by changing social relations with biophysical nature' (McCarthy and Prudham 2004: 277, 275, italics in the original), involving 'the privatisation and marketisation of ever more aspects of biophysical reality, with the state and civil society groups facilitating this and/or regulating only its worst consequences' (Castree 2008: 143). At the same time, environmental concerns 'represent the most powerful source of political opposition to neoliberalism' (McCarthy and Prudham 2004: 275) – an opposition which, as various chapters in this book highlight, finds an increasingly important terrain in technoscience. What remains to be seen – and this is a topic which also features prominently in the following pages – is the extent to which the 'neoliberalization of nature' represents a mere 'continuation of a more deeply historical process' (Heynen et al. 2007a: 10) or, through technoscientific advancement, introduces elements of real novelty into the relationship between biophysical nature and human agency. In other words, if neoliberalism can be considered as 'a particular historical variant of capitalism' (Heynen et al. 2007b: 287), what is the import of this variation, and how is technoscience embroiled in the process?

Any attempt to address this type of issue entails regarding neoliberalism as a particular way of responding to questions of who can govern, who and what is to be governed, and how. In this sense, from a Foucauldian perspective, neoliberalism embodies a specific 'governmentality'– a peculiar rationality or mentality of government (Foucault 2008). Key elements of this mentality are:

a) The market is the central institution of society, and a point of reference for handling any type of social affairs at any level (Foucault 2008, Feher 2009, Steger and Roy 2010). For example, it has been noted that 'neoliberal orthodoxies circulate [also] through and hybridize with environmentalism' (Heynen et al. 2007a: 11), triggering ambivalent or ambiguous forms of 'green governmentality' (Rutherford 2007). This entails a general downplaying of politics and political institutions in favour of technicized policymaking, based on benchmarking techniques and cost–benefit assessments (Hay 2007, Pestre 2009). It also entails the overarching relevance of a capitalist measurement of things (everything is 'capital': human, social, cultural, cognitive, biological), and a tendency towards growing abstraction, immateriality and temporal projection (speculation, hope, hype).
b) The market, however, is not a spontaneous institution but an artefact, something to be created and regulated purposefully (Tickell and Peck 2003). Hence, we have a prominent role of the rule of law, and the need to expand property rights over formerly unaffected social and biophysical sites.
c) The key figure of neoliberalism is the entrepreneur, and the key social mechanism is competition. As von Mises declared (see Peet 2007: 73), egoism is the basic law of society. At the same time, competition is to be promoted by enhancing the disparity of positions as much as possible, at the individual, corporate and state level (Ward and England 2007, Lazzarato

2009). Neoliberal rule is peculiarly focused on indirect, and often expert-mediated, forms of government aimed at informing and orienting the self-regulation of individual and corporate agents (Burchell 1996, Rose 1996a, Dean 1999). In this way, a circular reasoning is enforced, so that a reality is constructed and simultaneously presented as corresponding to the inherent properties of economic markets and human nature (Lemke 2003, Birch and Mykhnenko 2010b, Pellizzoni and Ylönen, this volume).

d) The primary goal of the market is the processing and conveyance of knowledge and information (Lave, Mirowski and Randalls 2010a): hence the profound interest of neoliberalism in ICTs (Harvey 2005), the irresistible expansion of financial markets, the performative role of technoscientific anticipations (Brown et al. 2000, Borup et al. 2006), and, more generally, the centrality of knowledge, communication and creativity for competition and profit (Cooper 2008, Suarez-Villa 2009). From this perspective, the market 'is not merely an economic phenomenon but also, and primarily, an epistemic one. […] A central concept of neoliberal political ideology is thus the "marketplace of ideas"' (Tyfield 2010: 63).

Overall, these aspects, which are variously addressed in the following chapters, show that, for the neoliberal mentality of government, the 'healthy' game of competition, and its legitimation as a 'natural' fact, cannot work if resources and rewards cannot be constantly changed or expanded, and vice versa. This brings advancements in science and technology to the forefront.

Neoliberalism and technoscience

The close connection between neoliberalism and technoscience is implicit in the latter's focus on supply-side economics, knowledge, innovation, and an intensified use of biophysical resources. Emblematic in this regard is the notion of 'Knowledge-Based Bio-Economy' (KBBE), which has been strongly promoted by think tanks and governmental institutions (see, for example, CEC-DG Research 2004, OECD 2009). According to the latter, economic growth and environmental sustainability, understood as increasingly eco-efficient industrial productivity, can be ensured by tightly coupling technoscientific knowledge and biological resources within a market framework. Sustainability, technoscience and the market, in other words, are seen as reciprocally strengthening each other. Market competition stimulates innovation, which enables growing eco-efficiency, which in turn triggers further market competition (Levidow et al., this volume).

As already noted, despite plenty of evidence of tight coupling, the literature which addresses the neoliberalism/technoscience relationship directly is rather limited, and mostly quite recent. Of course, the relationship is dealt with extensively in a loose sense, yet the apparent resistance to addressing the topic frontally is rather telling. The issue must reside in the aforementioned problems with the notion of neoliberalism; yet it also has partly to do with difficulties in connecting the two analytical spheres. The literature on neoliberalism, globalization or late capitalism does not seem particularly interested in drawing on the literature on technoscience and its most recent developments, and vice versa.

Take, for example, the area that Castree (2008) calls 'institutional political economy approach' and that can be related, directly or indirectly, to Marxian political economy. The crucial role of technoscience in the 'neoliberalization of nature' is often acknowledged, and yet there are no prominent attempts at elaborating on the specific features and performances of technoscience under neoliberal rule. If the problem with neoliberalism is in part, or maybe primarily, its implications for 'nature', there have been few efforts to reflect on the ways in which the very notion of nature is changing as a result of technoscientific advancement and reorganization within a neoliberal institutional and cultural framework (McCarthy and Prudham 2004, Harvey 2005, Castree 2008). The signs of growing interest in this issue are quite recent (see e.g. Birch and Mykhnenko 2010a), or may be found in non-mainstream sources of neo-Marxist thinking, like the philosophical reflections on late capitalism of Hardt and Negri, Paolo Virno and Maurizio Lazzarato, or else in sociological inquiries at the crossroads of a variety of intellectual streams, such as Suarez-Villa's (2009) study of 'technocapitalism', or Swyngedouw's (2010) discussion of climate change policies. Whether, and to what extent, the full implications of technoscientific change are captured by this literature is a question to which some pages of this book are devoted (see Gandini, Pellizzoni and Ylönen). Whatever the case may be, there seems to be plenty of work to be done in the neo-Marxist camp as far as the neoliberalism/technoscience relationship is concerned.

Even more elusive is the literature which has acquired a prominent role in thematizing late capitalism and environmental change, namely the so-called 'ecological modernization' approach (Mol, Sonnenfeld and Spaargaren 2009). Contrary to neo-Marxists, authors referring to this framework do not aim at major systemic changes, but rather at 'redirecting and transforming "free market Capitalism" in such a way that it less and less obstructs, and increasingly contributes to, the preservation of society's sustenance base' (Mol and Jänicke 2009: 24). Crucial to the reciprocal support of economic growth and environmental protection is technological innovation, and a political modernization focused on decentralizing, networking, adopting market-based and voluntary policy instruments, promoting responsible corporate and consumer self-regulation. Despite obvious similarities with neoliberal approaches (and some significant differences, such as the emphasis on non-contractual and non-territorial forms of 'ecological citizenship', and on excessive inequalities as a major source of problems), the latter slip into the background of most of the ecological modernization literature. At the same time, and with some exceptions (scholars dealing with environmental flows and related socio-technical networks), innovation and its role in social change are generally understood in a straightforward manner (Pellizzoni 2011a), with limited attention paid to conceptual and empirical insights from the field of science and technology studies (STS).

Also in this field, however, the situation is less than ideal. STS scholarship offers valuable analyses of the changes which have occurred in the past thirty years in the regulatory, institutional, financial, ideational, motivational and operational set-up of technoscience, from core research to the societal spread of applications. Attention has focused especially on biotechnologies and ICTs, and their interplay, with plenty of studies (indeed, too many to be listed here) on the informatization and commercialization of biomedicine and biotechnology, and on the reorganization of academic research. More recently, a growing body of literature has centred on nanotechnologies, 'converging technologies' and related policy and R&D narratives and programs (Roco and Bainbridge 2002, Nordmann 2004; see also Arnaldi this volume, Bárd this volume), as well as on the role of hype and expectations in orienting research (Brown and Michael 2003, Borup et al. 2006, Pollock and Williams 2010). To date, however, STS have had relatively little to say about the changes in the science–society relationship associated with neoliberalization (Moore et al. 2011). The existing literature 'has been notably vague about neoliberalism's definition and temporal and geographical specificities, as well as the extent to which neoliberal political–economic relations beyond academia shape what happens within it, […] [possibly as] one legacy of the modern STS field's repudiation of the Mertonian division between internal and external influences on science' (Lave, Mirowski and Randalls 2010a: 660). Many works, remaining faithful to an established tradition that privileges micro-empiricist approaches, relegate the neoliberal framework to a vague, and often implicit, background. For example, recent assessments of the relationship among genomics, nature, environment and society (Parry and Dupré 2010; Carter and Charles 2010) mostly avoid venturing into the properly political dimensions of the issue, and largely steer clear of neoliberalism as a concept. Yet studies that take a broader, institutional perspective, addressing issues of science/government, science/industry and science/social movements relationships, also refrain from dealing with neoliberalism (or late capitalism), preferring (allegedly) less normatively or politically charged notions, such as globalization, knowledge society, civic epistemology, socio-technical imageries and so on (see e.g. Nowotny, Scott and Gibbon 2001; Jasanoff 2005; Jasanoff and Kim 2009; Felt and Wynne 2007). There are also opposite signs, as shown, for example, by the collection of essays edited by Szerszynski and Urry (2010). This collection, however – at least in our view – belongs more to environmental sociology than STS. Indeed, a number of interesting, albeit often indirect, accounts of the interplay of neoliberalism and technoscience can be found in the environmental sociology area.

Complaining about the insufficiency of current STS scholarship, because of its focus on the internal changes of science to the detriment of external pressures and its concern for biotechnology and biomedicine to the detriment of other natural and social science areas, Lave, Mirowski and Randalls (2010a) make a case for paying specific attention to alterations in the institutional set-up and functioning of science induced from the outside. The institutional neoliberalization of science includes a number of interconnected aspects: a rollback of government support to, and organization of, public research universities, replaced by increased corporate funding (rather than in-house research, firms prefer to invest in targeted research by contract research organizations or university-based scientists); a reframing of universities' mission as providers of human capital and competitive global service industries rather than educational institutions, with the consequent expansion of non-tenured and post-doctoral positions; a growth of industries devoted to the ghost-writing of papers and the ghost management of research (buying research increasingly includes 'buying' the persons to whom such research will be attributed); and the aggressive promotion and protection of intellectual property: that is, the commodification of knowledge (and to some extent of ideas themselves), despite the dubious financial advantages for the vast majority of universities. In synthesis, 'neoliberal science policy is creating a regime of science organization quite distinct from the Cold War science management regime' (Lave, Mirowski and Randalls 2010a: 667). More precisely, one might say, the rise of neoliberalism 'intensifies' (Nealon 2008) already existing connections between scientific and commercial purposes, engendering a qualitative change in science both as institution and practice.

In short, Lave, Mirowski and Randalls (2010a) make a strong case for complementing the currently predominant sociology of scientific knowledge (SSK) with robust injections of an institutional sociology of science. In this way, however, the focus is entirely on the 'neoliberalization of science': the new organization entailed by neoliberal reforms; the 'relations between private profit and public science' (Lave, Mirowski and Randalls 2010a: 660); the new ethos of research, symbolized by the figure of the scientist-entrepreneur and the start-up company. What remains in the shadows is the symmetrical process of scientization of neoliberal governance. The problem is not only 'how neoliberal theories of society are transforming technoscience' (Lave, Mirowski and Randalls 2010a: 669), but also how technoscience contributes to the neoliberalization of society.

In a recent book, Andrew Feenberg (2010) evokes a famous work of Maurits Escher, entitled 'Drawing Hands' (two hands, drawing each other, emerge from the paper). As often happens with Escher, it is impossible to establish an order or ranking of events, to say which element of the drawing is the origin or determinant of the other, which is the subject and which is the object. Feenberg refers to this image to depict the co-production of society and technoscience. Like a number of other contributions that build on the idea of science-society co-production (e.g. Nowotny, Scott and Gibbon 2001, Jasanoff 2004 and 2005) – and despite dealing extensively with globalization, power, freedom, critique and cultural change vis-à-vis technology – Feenberg's book pays only marginal attention to neoliberalism (the index shows just one quotation in more than 250 pages). Escher's work, however, can portray the topic and approach of this collection of essays just as well. Neoliberalism and technoscience, we believe, develop in close connection, and profoundly affect each other. As Antoinette Rouvroy remarks in her study of the relationship between genetics and neoliberalism, 'the social/economic/technical/political structure of society and innovation [are] related to each other, as part of the same metabolism, interacting in a dialectical manner, each being performative for the others' (Rouvroy 2008: 6).

To grasp this mutual constitution, a 'political sociology of science' (Frickel and Moore 2006) is arguably required as a complement to Mertonian and SSK approaches. This perspective regards both neoliberalism and technoscience as quasi-autonomous social fields, partly working according to their own logics and processes, and partly subject to reciprocal influences, in terms of countervailing tendencies and the intertwining of scientific and political languages and rationales that stabilize and modify institutional relationships. This perspective, therefore, highlights not only that academe has undertaken a profound restructuring, but also that public regulation has undergone a process of scientization – with a growing role of corporate self-governance, transnational technical bodies and a technical governance of allegedly politically neutral issues of safety, quality and efficacy of innovation – and that civil society and social movements are playing an increasingly active and multifaceted role in the interpenetration of neoliberalism and technoscience (Moore et al. 2011, Hess, this volume; see also Barry 2002, McCormick 2007, Felt and Wynne 2007, Kinchy, Kleinman and Autry 2008, Hess 2009, Frickel et al. 2010, Ottinger 2010, Pellizzoni 2011b, Blok 2011 and this volume).

A renewed institutional approach to technoscience, it is argued, benefits from a political, economic and organizational sociology insight (Moore et al. 2011). We agree, although we also believe that it is important to retain a profound lesson of SSK, namely that science and society (politics, economy, culture) exert a reciprocal influence not only at an institutional or organizational level, but also, and perhaps primarily, at the level of their rationalities – the ways in which action is selectively oriented in each domain according to meanings, assumptions and goals that draw from other domains. In other words, the relationship or reciprocal constitution of neoliberalism and technoscience is a multilayered affair which traverses all levels of social life, from the more intimate processes of sense-making to the larger institutional manifestations – often physically tangible, such as the 'science parks' which have flourished across the world to host 'innovation incubators' (CEC-DG Regional Policy 2010), and other activities which connect innovation and entrepreneurship. Carbon trading is an excellent example of this. It builds on and testifies to the overlap between an understanding of the market as an information processor, and of scientific knowledge as a resource whose abstracting capacities can be capitalized, translating physical entities and processes into tradable commodities (MacKenzie 2009, Lohmann 2010, Blok, this volume). More than the institutional set-up of science and politics, what is at stake here is their intimate reason, their understanding of the knowledge/power relationship between the biophysical world and human agency (Pellizzoni 2011a).

This remark, by any standards, calls into question the governmentality perspective. And indeed, it is within the governmentality literature that one finds a sustained treatment of the relationship of neoliberalism with technoscience. This is hardly surprising. First, the link between science and politics has been strongly stressed by Foucault. More precisely, for him (Foucault 2003: 239 ff.) the 'biopolitical' technologies of government which emerged in the late eighteenth and developed in the nineteenth and twentieth centuries – that is, together with the spread of classical and social liberalism as political-economic doctrines and ways of organizing the state – are characterized by a focus on the 'population' (or the human species) in four fields: natality, morbidity, inabilities, and the relationship between humans as living beings and their surrounding environment. All these areas of intervention trigger the development of a variety of disciplines, expertise, and organizational and technological devices. Secondly, Foucault has explored the character of neoliberalism, in terms of both continuity with and change from liberalism, providing insights today subject to intense debate (see e.g. Nealon 2008, Venn and Terranova 2009, Binkley and Capetillo 2009, Tazzioli 2011). Moreover, the Anglo-American Foucauldian school – Nikolas Rose, Paul Rabinow, Peter Miller, Mitchell Dean, Carlos Novas and others – has become highly influential in the study of the societal impacts of a technoscience that appears to be heavily affected by neoliberalism's ideational, regulatory and institutional aspects. Biotechnologies, health and security are predictably prominent as topics here, and attention is focused on the micro level of power relations, with special reference to the impact of neoliberalism on individuals' self-understanding as ethical subjects – hence a marked interest in the role of the social and neuro-sciences (Rose 1996b, Maasen and Sutter 2007, Arnaldi, this volume, Bárd this volume, Ferreira et al. this volume). Overall, this literature offers a nuanced, controversial picture of the intertwining of neoliberalism and technoscience. The very term neoliberalism (or 'advanced liberalism') that featured extensively in earlier works (Barry, Osborne and Rose 1996) tends to disappear from recent studies (Rose 2007). A debate has arisen (Myers 2008, Collier 2009, Raman and Tutton 2010) around the contentious meaning of Foucault's own development of the notion of 'care of the self'; a debate which has been transferred to the interpretation of neoliberal biopolitics as entailing either overcoming the traditional normalizing strategies targeted on the control of bodies and populations, in favour of new securitarian strategies grounded in personalized techniques of government at a distance or self-regulation, or a combination of the two, together with constant recourse to the 'old' sovereign power. One clue to the profound interpretive divide on these processes – one typical subject matter is genetic screening – is the contrast between those who insist on, or on the contrary contest the possibility of, distinguishing between a 'subjecting' *biopower* and an 'emancipatory' or 'subjectivating' *biopolitics* (Hardt and Negri 2004, Rose 2007, Esposito 2008, Marzocca 2007). The most critical, unsurprisingly, are those who combine the governmentality approach with a Marxian insight (Lemke 2003, 2004; Lohmann 2010; Swyngedouw 2010). A pivotal concept in this area of debate is that of 'biocapital' (or 'biovalue'). This notion is regarded as crucial to exploration of 'the relationship between materiality and modes of abstraction that underlie the co-emergences of new forms of life science with market regimes for the conduct of such science [...] [on the grounds that] life sciences represent a new face, and a new phase, of capitalism and, consequently, that biotechnology is a form of enterprise inextricable from contemporary capitalism' (Rajan 2006: 33, 3). Rajan seeks to combine STS ethnographic methods and the STS-sourced idea of co-production or co-evolution of the scientific and the social fields, with a Marxian outlook on the current forms of global capitalism, and a Foucauldian understanding of the institutional, epistemic and discursive mechanisms of power. A comparable effort is made by Cooper (2008), whose analysis of the relationship among politics, economics, science, and cultural values in the US elaborates on the notion of life (especially human life) as surplus, making a strong case for a simultaneous account of the rise of biotechnology and of neoliberalism as a political force and an economic policy.

Despite the relevance of these studies, however, it is important to remember that the interaction of neoliberalism and technoscience does not only take place in the field of biotechnologies and biomedicine; as already seen, it extends on the one hand to the linguistic, symbolic and communicative capacities of humans, and on the other, to 'nature' or 'the environment' as a whole (Lazzarato 1997 and 2004, Suarez-Villa 2009, Virno 2004 and 2009, Ferreira et al. this volume, Gandini this volume).

About this book

To sum up, as a programme of economic-political change, and even more so as an all-encompassing – if patchy – design of social transformation, neoliberalism involves science as a major object and trigger. A thrust towards renewed accumulation processes entails an extension and intensification of access to, and appropriation of, natural resources. A supply side economics centred on competition relies crucially on technoscientific advancement and innovation. The latter becomes a major site of entrepreneurship while, conversely, scientific inquiry takes a markedly entrepreneurial direction. Market competition becomes the basic form of regulation of technoscience. The market as a whole centres increasingly on knowledge and information. This pervasive mentality of government penetrates the logic of research, its motivations and underlying assumptions about the features of the natural and social world, and what one can and may want to do with it.

The relationship between neoliberalism and technoscience is thus a complex and important question. Making sense of, and analysing, its multifarious aspects represents a major challenge. The issue has been addressed in many ways, but certainly needs further, systematic research. As has been seen, there are indications of a growing interest in a focused inquiry (Birch and Mykhnenko 2010a, Lave, Mirowski and Randalls 2010b, Szerszynski and Urry 2010, Moore et al. 2011). A book explicitly devoted to assessment of the theme of neoliberalism and technoscience therefore seems to be a timely endeavour: all the more so if it arrives at a moment when neoliberalization processes are undergoing major reconfiguration. Even if arguments about 'after neoliberalism', 'post-neoliberalization', or the exit from neoliberalism towards new, more socially-oriented forms of global governance, are being advanced not only at the social movement level but also in academic debate (see e.g. Brand and Sekler 2009), we believe that the social, cultural and institutional transformations occasioned by neoliberalism are bound to have an enduring legacy.

The starting point of this book, as said, is that technoscience is crucial to, and crucially affected by, the neoliberalization of society in all its aspects (political, economic, institutional, and cultural), and at all levels of agency (governmental, corporate, associational, and individual). The preceding outline has indicated a number of conceptual, methodological and substantive perspectives which might be relevant to treatment of this topic. It is of course impossible, and probably also undesirable, to account fully for such variety in one single book. Yet by combining a range of theoretical and empirical insights, this collection of essays aims to strike a balance between richness and focus, breadth and consistency.

The book brings together scholars from different countries and generations, which is just as well in our view because interests, perspectives and sensitivities are also a matter of education, professional trajectories and research experiences. Some features also characterize its content. To begin with, there is no single theoretical framework. However elegant this solution might seem, we believe that in the present state of inquiry it would turn out to be forced, and would limit, rather than strengthen, the analytical insights. The chapters draw on, and often combine, Marxian political economy and radical neo-Marxism, governmentality and actor-network theory, as well as other perspectives, such as Boltanski and Thévenot's sociology of critique, political theory and organizational analysis. A second feature is that biocapital, bioeconomy and biotechnology feature substantially, but not exclusively, as fields of inquiry. Other issues such as ICTs, climate change regulation and psychiatry, are addressed as well. Thirdly, though drawing to various extents on historical insights, the book as a whole is mainly oriented towards questioning the present (and prospective) state of affairs. Finally, compared with the micro-empiricist orientation which is fairly common in the STS field (but examples of which can also be found in the literature which focuses on the neoliberalization of space, territory and nature: see e.g. Heynen et al. 2007c), the following chapters take a different type of approach, settling mostly at the level of intellectual perspectives, disciplinary trajectories, policy fields, regulatory solutions, and related struggles.

The book is divided into three parts, each including three chapters. Given the intertwining of themes and issues, there are no really compelling arguments in support of this kind of division and assignment of chapters, but the way the book is structured reflects what seem to us to be loosely distinguishable areas of inquiry.

The first part is entitled 'Neoliberalism, technoscience and late capitalism'. Here, we have collected chapters which provide different sorts of encompassing outlook on the institutional, governmental and cultural aspects of the neoliberalism–technoscience relationship.

In Chapter 1, 'Neoliberalism and technology: Perpetual innovation or perpetual crisis?', Laurence Reynolds and Bronislaw Szerszynski approach the relationship between neoliberalism and technoscience from a historical perspective, questioning the capacity of capitalism to overcome its resurgent crisis by resorting to new forces of production. The period of growth after the Second World War was characterized by the Fordist enterprise, an interventionist state and a set of productive forces based on continuous technological innovation. This trend seemed to intensify with the onset of the crisis around 1970, thanks to new forces of production based especially on information technology and biotechnology. By applying a Marxian framework the authors argue, however, that neoliberalism and its related suites of technologies have not properly addressed the crisis of Fordist capitalism. The expansion in the 1990s and after is misleading in both its character and source. Profit has been generated more through promissory speculation than commodity production. The crucial innovation focuses on the respatialization of productive forces around the world, thanks to mundane technologies such as diesel engines, containers, ships and logistical information technologies, or low-paid work in traditionally organized factories. This contradicts much-publicized allegations of a 'third industrial revolution', instead indicating fundamental stagnation. Yet the neoliberal period has witnessed a major change in the relations between science and capitalism. More than a mere privatization, science has been subject to a 'secondary primitive accumulation', by which value is extracted from intellectual resources that remain collective and non-commodified. In short, neoliberalism involves a low-tech economic 'revolution' which fails to engage science fully as a productive force, and simultaneously a high-tech science that is cannibalized or 'harvested' as a commodity.

In Chapter 2, 'Hegemonic contingencies: Neoliberalized technoscience and neorationality', we elaborate on two different, but to some extent complementary, theoretical insights – Foucauldian governmentality and Marxian dialectical materialism, with special reference to hegemony theory – to shed light on what we depict as the specific rationality of government which underlies the reciprocal constitution of neoliberalism and technoscience. We argue, more precisely, that the hegemonic status of the former draws significantly on this rationality, as enacted in and by the latter. This novel form of rationality stems from the intensification of contrasting traits of the modern cultural heritage, turning out to be hyper-modernist and post-modernist at the same time. This 'neorationality', by which a strong human agency is complemented by a full pliancy of nature, translates a mutually constitutive interaction, as described in much STS literature, into a matter of appropriative control. At the same time, the oscillation of neoliberal discourses between description and prescription makes it particularly difficult to develop a critique. Technoscience is profoundly traversed by this logic, leading to a peculiar way of conceiving and handling contingency and instability. Biotechnology regulation and practices offer a paradigmatic example of the simultaneous neoliberal remoulding of nature, people and politics. Moreover, this very sector shows how the development of governmental devices grounded on 'ethics' and 'dialogue' acts as a major trigger for the neoliberalization of technoscience. The chapter concludes with reconsideration of the question of 'resistance' in the light of the peculiarly hegemonic character of neoliberal rationality.

In Chapter 3, 'Neoliberalism and ICTs: Late capitalism and technoscience in cultural perspective', Alessandro Gandini addresses the role which information technologies play in the intertwining of neoliberalization processes and technoscientific evolution, arguing that a cultural perspective is the key to understanding the encounter between technoscience and neoliberalism in the context of late capitalism. ICTs have represented a condition for the emergence of neoliberalism as a political-economic vision, yet they are now nourishing opportunities for its overcoming. The trajectories of ICTs and neoliberalism, in other words, are closely connected but tendentially diverging. Gandini draws his argument from classic analyses and recent insights into the cultural logic of late capitalism. The conceptual equipment for interpreting the role of knowledge and creativity in current technocapitalism is provided by the works of Hardt and Negri as well as of Lazzarato and Virno. ICTs dramatically affect the capitalist mode of production. They are at the centre of cognitive capitalism. The latter focuses on the extraction of value from the immaterial, cognitive competencies of intellectual workers, yet it is unable to affect the cultural and social dynamics developing among them according to the space of possibilities offered by communications. ICTs produce a collective subject, the 'multitude', which, as an heterogeneous assemblage of singularities, has hardly anything in common with the collectives which are typical of modernity (people or class); this is a subject within which new forms of sociality, new economic visions and new solidarity networks emerge, and which points to a direction very different from the neoliberal model of economy and subjectivity.

The remaining two parts of the book address the intertwining of neoliberalism and technoscience as bound up with two major fields of action: humanity, on the one hand, and the natural environment on the other.

The second part opens with Chapter 4, 'The end of history and the search for perfection. Conflicting teleology of transhumanism and (neo)liberal democracy'. Here, Simone Arnaldi elaborates on one of the most prominent features of the neoliberalism/technoscience relationship, namely its engagement with promissory anticipations of technological futures: an engagement which, as the chapter shows, is far from stable and consistent. Arnaldi addresses this topic from the vantage point of the connections between what he labels 'a problematic triad': neoliberalism, as a specific regime of technological innovation; transhumanism (understood as being synonymous with posthumanism), as an emergent movement promoting an alleged technology-enabled overcoming of human mental and physical limits; and the position of one of the most influential champions of liberal democracy and adversaries of the 'human enhancement' argument: Francis Fukuyama. More than speculation, neoliberalism, transhumanism and Fukuyama define a *telos*, a perfected state for human society to be actively pursued, with convergences and divergences in their arguments. Notions of selfhood and market join neoliberal and transhumanist discourses together, while Fukuyama builds his counterargument on his own liberal philosophy of history, according to which posthumanism represents a threat to the natural evolution of human political history. The controversy over human enhancement corresponds to a conflict of teleologies which revolves around different notions of 'perfection' and 'perfectibility' – the subject of perfectibility, the limits to the perfecting action, and the time when perfection is reached. Divergences, however, are fuelled by similarities. All three positions are strongly normative – no alternative state of affairs is admitted in terms of human development. Moreover, the future is assumed to be disclosed to present knowledge, pre-empting discussion on technological choices.

Chapter 5, 'Pre-empting the threat of human deficiencies', nicely complements this discussion. For Imre Bárd enhancement measures are becoming preventive and pre-emptive interventions. Expanding and optimizing vitality is a crucial element of neoliberal governmentality. The earlier focus on populations is replaced by a regime of individualized enhancement, and the economic target of increased productivity is intertwined with self-technologies aimed at securing, optimizing and improving our biological constitution, our personal health and well-being, in a prudent, responsible and calculated manner. On closer inspection, however, the shift from population- to individual-centred biopolitics is far from unquestioned. Biological technosciences dissolve the individual into bodily markers, neurotransmitter activity, biobank data, and probabilistic or tendential biological processes. At the same time, there is a resurfacing of the population as a guiding notion for policy action. A case has been made for submitting populations to enhancement measures and selection techniques in order to survive environmental and social challenges. Within bioethics, too, the individual seems to be losing his/her privileged position in favour of more communitarian principles. In short, human enhancement is gradually reconstituted along the lines of preventive measures that no longer serve the purposes of fulfilling individual desire or furthering choice – as the neoliberal claim would have it – but are aligned with a logic of risk prevention in a world suffused by anxiety, by the dread of external and internal threats. The spread of discourses on flaws and potential dangers testifies to this drift in the form, for example, of forensic genetic databases and algorithms for filtering biological and behavioural information. In short, the constitution of the human being is seen as inherently deficient and in need of improvement, while human deficiencies are simultaneously framed as threats. In this way, enhancement becomes a form of prevention aimed at managing the risks of being human.

A third perspective on how humanity is embroiled in the neoliberalism/technoscience relationship is offered by Chapter 6: 'The question of citizenship and freedom in the psychiatric reform process: A possible presence of neoliberal governance practices'. Arthur Arruda Leal Ferreira, Karina Lopes Padilha, Míriam Starosky and Rodrigo Costa Nascimento dwell on the long-recognized role of the *psy*-sciences from a particular viewpoint: the ambiguous implications of the reform of psychiatric treatment inaugurated in Italy between the 1960s and 1970s, and in Brazil in the 1980s. This was an 'anti-asylum', 'emancipatory' reform aimed at breaking with the traditional, disciplinary approaches to mental illness through the development of 'open door' institutions and social integration programs. The theoretical background of the chapter, which draws on laws, official documents, academic texts and concrete practices, is Foucault and the Foucauldian school. Yet this perspective is complemented with STS insights, namely Latour's concept of 'circulatory system', to account for the status of the *psy*-sciences as a heterogeneous assemblage of frameworks and practices. The establishment of psychology and psychiatry is closely linked to the intertwining of – more than the transition between – sovereign and disciplinary powers on the one hand, and the liberal techniques of government based on self-regulation and scientific knowledge on the other. The authors argue that techniques to manage people are present not only in the old psychiatry but also in the new reformist process. The latter, as a consequence, runs the risk of inadvertently endorsing a neoliberal 'self-enterprising' understanding of freedom and autonomy. To counter this risk, psychiatric reformers should pay especial attention to the way in which social inclusion is concretely enacted through work, consumption, political participation and self-investment.

As said, the third part of the book addresses the relationship between neoliberalism and technoscience from the vantage point of its environmental implications. The title of Chapter 7, 'Neoliberalism, technoscience and environment: EU policy for competitive, sustainable biofuels', effectively summarizes its content. After a theoretical discussion on the connections among market, environment and technoscience under neoliberal rule, Les Levidow, Theo Papaioannou and Kean Birch focus on the EU biofuels policy as an emblematic case of neoliberal 'green governmentality', where a virtuous circle is assumed to exist among technoscientific innovation, market exchanges, and answers to social needs. The evolution of biofuels policy also offers a typical example of technological controversy. A certain strategy is decided, in this case for a 'renewable energy' policy aimed at responding to a variety of issues: energy security, greenhouse gas savings, technology export and rural development. A controversy then explodes over the harmful environmental and social effects of this policy, especially in the global South, where land-use changes anticipate and supply a larger EU market for biofuels. Experts and policymakers then reply to criticisms by drawing a distinction between problem-raising 'conventional' biofuels and solution-providing new 'sustainable' biofuels. Appropriate incentives and standards are expected to engender technoscientific innovation towards a more efficient use of renewable resources, leading to more competitive and sustainable biofuels. A promissory orientation towards technoscientific fixes is framed in terms of market opportunities and economic imaginaries. This strengthens the link among markets, technoscience and the environment. The key idea of exploiting 'marginal land' is emblematic of the process of 'accumulation by dispossession' whereby the social-biophysical world is reshaped as a commodity suited to the logic of global value chains and capital accumulation, regardless of the environmentally destructive and socially exclusionary implications.

In Chapter 8, ‘Configuring homo carbonomicus: Carbon markets, calculative techniques, and the green neoliberal’, Anders Blok sheds light on another site of entanglement among neoliberalism, technoscience and the environment. Combining the governmentality approach with insights from actor-network theory and Boltanski and Thévenot’s sociology of critique and justification, Blok furnishes a nuanced account of a much-debated topic. The rise of transnational markets for trading in CO_2 emissions constitutes an emblematic example of the growing importance of neoliberal forms of environmental governance. Putting a price on emissions is assumed to promote industrial energy efficiency, ‘green’ technological innovation, and the transfer of low-carbon technology to developing countries. The result is a multi-billion dollar economy based on carbon finance, green-tech firms, project developers, and consultants. The chapter explores the way in which carbon has been forged as a neoliberal object, linking together technoscience, economics, politics, public commitments, calculations and normative investments, policy choices and everyday life, to produce a particular form of human agency which Blok calls ‘homo carbonomicus’. A careful examination of this figure shows its hybrid character, at the crossroads among nature commodification, technoscientific debate and ethical-political concern. Carbon markets cannot sustain themselves by market commitments alone. They combine a variety of regimes of worth, which include technical, civic and green values. This means that market-based regulation is anything but a depoliticized field. On the contrary, technoscientific and political ambiguities open up spaces for public protest, monitoring and re-orientation – as testified by the differentiated strategies adopted in this field by environmental NGOs.

In Chapter 9, ‘The green transition, neoliberalism and the technosciences’, David Hess takes a political sociology approach. As we have already remarked, this perspective combines cultural, class and institutional analyses, and it entails a notion of neoliberalism and technoscience as quasi-autonomous, but interrelated, fields of practices and coalitions. Whilst in previous works Hess has identified three primary areas of inquiry – the transformation of academic work, civil society’s politicization of research agendas, and changes in the role and regulatory approach of the state – the essay included in this collection addresses only the last issue, focusing on trade and industrial policy in the US as it relates to green energy technology. Hess argues that, rather than being a totalizing regime, neoliberalism is just one among several different contending ideologies embedded in specific policy orientations. Neoliberalism is one type of liberalism, which maintains an enduring tension with the social liberal, or social democratic, variant of liberalism as regards the role of the state in the economy; a tension that punctuates American history in the twentieth and early twenty-first centuries. Other, less prominent, positions are socialism and localism and, with special reference to the green energy policy, what Hess calls ‘developmentalism’. These orientations take different stances vis-à-vis trade liberalization; resource distribution; the role of public and private, and central and local, actors; and diverse technological and organizational designs. Caution is therefore required as regards the alleged all-encompassing relationships between technoscience and neoliberalism. Scientific research programmes, technological solutions and choices, and political ideologies are linked in a contingent way. Making sense of socio-technical change, in other words, requires a fine-grained, historically sensitive, type of analysis.

The concluding chapter highlights what in our view are the main insights offered by the book. The intimate, reciprocally constitutive relationship of neoliberalism and technoscience emerges in all its intricacy, as testified by some tensions emerging from the chapters: about the elements of continuity and change of neoliberalism vis-à-vis previous forms of liberalism; the compromises and contradictions of neoliberal policies that different technoscientific fields highlight; the surfacing or likely outcomes of such ambivalences or contradictions.

The book has its origins in a group of papers on neoliberalism and technoscience discussed at the conference of the European Association for the Study of Science and Technology (EASST) held in Trento in September 2010. We are grateful to Ashgate Publishing, and to Neil Jordan in particular, for suggesting and supporting the development of a collection of essays, and to the authors, who were present at the conference or who subsequently teamed up, for accepting the challenge of confrontation with a complex and contested, but also important and fascinating, issue.

References

Arrighi, G. and Silver, B. 2001. Capitalism and world (dis)order. *Review of International Studies*, 27(5), 257–79.

Barnett, C. 2005. The consolations of 'neoliberalism'. *Geoforum* 36(1), 7–12.

Barry, A. 2002. The anti-political economy. *Economy & Society* 31(2), 268–84.

Barry, A., Osborne, T., Rose, N. 1996. (Eds.) *Foucault and Political Reason*. London: UCL.

Binkley, S. and Capetillo, J. 2009. (Eds.) *A Foucault for the 21st Century: Governmentality, Biopolitics and Discipline in the New Millennium*. Newcastle: Cambridge Scholars Publishing.

Birch, K. and Mykhnenko, V. 2009. Varieties of neoliberalism? *Journal of Economic Geography*, 9(3), 355–80.

Birch, K., Mykhnenko, V. 2010a. (Eds.) *The Rise and Fall of Neo-Liberalism*. London: Zed Books.

Birch, K., Mykhnenko, V. 2010b. Introduction: A world turned right way up, in *The Rise and Fall of Neo-Liberalism*, edited by K. Birch and V. Mykhnenko. London: Zed Books, 1–20.

Blok, A. 2011. Clash of the eco-sciences: Carbon marketization, environmental NGOs, and performativity as politics. *Economy and Society*, 40(3), 451–76.

Borup, M., Brown, N., Konrad, K. and Van Lente, H. 2006. The sociology of expectations in science and technology. *Technology Analysis and Strategic Management*, 18(3/4), 285–98.

Bourdieu, P. and Wacquant, L. 2001. New Liberal speak: Notes on the new planetary vulgate. *Radical Philosophy* 105.

Brand, U. and Sekler, N. 2009. (Eds.) Postneoliberalism – A beginning debate. *Development Dialogue*, 51, special issue.

Brenner, N. and Theodore, N. 2002. Cities and the geographies of 'actually existing neoliberalism'. *Antipode*, 34(3), 349–79.

Brenner, N. Peck, J., Theodore, N. 2010. Variegated neoliberalization: Geographies, modalities, pathways. *Global Networks*, 10(2), 182–222.

Brown, N. and Michael, M. 2003. A sociology of expectations: Retrospecting prospects and prospecting retrospects. *Technology Analysis & Strategic Management*, 15(1), 3–18.

Brown, N., Rappert, B. and Webster, A. (eds.) 2000. *Contested Futures. A Sociology of Prospective Techno-science*. Aldershot: Ashgate.

Brown, W. 2006. American nightmare. Neoliberalism, neoconservatism, and de-democratization. *Political Theory*, 34(6), 690–714.

Bumpus, A. and Liverman, D. 2008. Accumulation by decarbonisation and the governance of carbon offsets. *Economic Geography*, 84(2), 127–55.

Burchell, G. 1996. Liberal government and techniques of the self, in *Foucault and Political Reason*, edited by A. Barry, T. Osborne and N. Rose. London: UCL Press, 19–36.

Campbell, J. and Pedersen, O. 2001. Introduction. The rise of neoliberalism and institutional analysis, in *The Rise of Neoliberalism and Institutional Analysis*, edited by J. Campbell and O. Pedersen. Princeton, NJ: Princeton University Press, 1–23.

Carter, B. and Charles, N. 2010. (Eds.) *Nature, Society and Environmental Crisis*. London: Wiley.

Castree, N. 2008. Neoliberalising nature: The logics of deregulation and reregulation. *Environment and Planning A*, 40, 131–52.

CEC-DG Research 2004. *Towards a European Knowledge-Based Bioeconomy*. European Commission, Directorate-General for Research. Luxembourg: Office for Official Publications of the European Communities.

CEC-DG Regional Policy 2010. *The Smart Guide to Innovation-Based Incubators (IBI)*. European Commission, Directorate-General for Regional Policy. Luxembourg: Publications Office of the European Union.

Collier, S. 2009. Topologies of power: Foucault's analysis of political government beyond 'governmentality'. *Theory Culture & Society*, 26(6), 78–108.

Cooper, M. 2008. *Life As Surplus: Biotechnology and Capitalism in the Neoliberal Era*. Washington: University of Washington Press.

Crouch, C. 2011. *The Strange Non-Death of Neoliberalism*. Cambridge: Polity Press.

Dean, M. 1999. *Governmentality*. London: Sage.

England, K. and Ward, K. 2007. (Eds.) *Neoliberalization: States, Networks, Peoples*. Oxford: Blackwell.

Esposito, R. 2008. *Bios. Biopolitics and Philosophy*. Minneapolis: University of Minnesota Press.

Feenberg, A. 2010. *Between Reason and Experience. Essays in Technology and Modernity*. Cambridge, MA: MIT Press.

Feher, M. 2009. Self-appreciation: Or, the aspirations of human capital. *Public Culture*, 21(1), 21–41.

Felt, U. and Wynne, B. 2007. (Eds.) *Taking European Knowledge Society Seriously*. Report for the European Commission. Luxembourg: Office for Official Publications of the European Communities. Available at: http://ec.europa.eu/research/science-society/document_library/pdf_06/european-knowledge-society_en.pdf [accessed: 5 June 2012].

Foucault, M. 2003. *Society Must Be Defended*. New York: Picador.

Foucault, M. 2008. *The Birth of Biopolitics*. London: Palgrave MacMillan.

Frickel, S. and Moore, K. 2006. (Eds.) *The New Political Sociology of Science*. Madison, WI: University of Wisconsin Press.

Frickel, S., Gibbon, S., Howard, J., Kempner, J., Ottinger, G. and Hess, D. 2010. Undone science: Charting social movement and civil society challenges to research agenda settings. *Science, Technology, and Human Values*, 35(4), 444–73.

Hardt, M. and Negri, A. 2004. *Multitude: War and Democracy in the Age of Empire*. New York: Penguin Press.

Harvey, D. 2003. *The New Imperialism*. Oxford: Oxford University Press.

Harvey, D. 2005. *A Short History of Neoliberalism*. Oxford: Oxford University Press.

Hay, C. 2007. *Why We Hate Politics*. Cambridge: Polity Press.

Hess, D. 2009. The potentials and limitations of civil society research: Getting undone science done. *Sociological Inquiry*, 79(3), 306–27.

Heynen, N., McCarthy, J., Prudham, S. and Robbins, P. 2007a. Introduction: False promises, in *Neoliberal Environments: False Promises and Unnatural Consequences*, edited by N. Heynen, J. McCarthy, S. Prudham and P. Robbins. London: Routledge, 1–21.

Heynen, N., McCarthy, J., Prudham, S. and Robbins, P. 2007b. Conclusion. Unnatural consequences, in *Neoliberal Environments: False Promises and Unnatural Consequences*, edited by N. Heynen, J. McCarthy, S. Prudham and P. Robbins. London: Routledge, 287–91.

Heynen, N., McCarthy, J., Prudham, S. and Robbins, P. 2007c. (Eds.) *Neoliberal Environments: False Promises and Unnatural Consequences*. London: Routledge.

Jasanoff, S. 2004. The idiom of co-production, in *States of Knowledge. The Co-Production of Science and Social Order*, edited by S. Jasanoff. London: Routledge, 1–12.

Jasanoff, S. 2005. *Designs on Nature. Science and Democracy in Europe and the United States*. Princeton, NJ: Princeton University Press.

Jasanoff, S. and Sang-Hyun, K. 2009. Containing the atom: Sociotechnical imaginaries and nuclear power in the United States and South Korea. *Minerva*, 47(2), 119–46.

Jessop, B. 2002. Liberalism, neoliberalism, and urban governance: A state-theoretical perspective. *Antipode*, 34(3), 452–72.

Kinchy, A., Kleinman, D. and Autry, R. 2008. Against free markets, against science? Regulating the socio-economic effects of biotechnology. *Rural Sociology*, 73(2), 147–79.

Lave, R., Mirowski, P. and Randalls, S. 2010a. Introduction. STS and neoliberal science. *Social Studies of Science*, 40(5), 659–75.

Lave, R., Mirowski, P. and Randalls, S. 2010b. (Eds.) STS and neoliberal science. *Social Studies of Science*, 40(5), 659–91.

Lazzarato, M. 1997. *Lavoro immateriale*. Verona: Ombre Corte.

Lazzarato, M. 2004. *Les révolutions du capitalisme*. Paris: Les Empêcheurs de Penser en Rond.

Lazzarato, M. 2009. Neoliberalism in action. Inequality, insecurity and the reconstitution of the social. *Theory, Culture & Society*, 26(6), 109–33.

Lemke, T. 2003. Foucault, governmentality and critique. *Rethinking Marxism*, 14(3), 49–64.

Lemke, T. 2004. 'Disposition and determinism – genetic diagnostics in risk society'. *Sociological Review*, 52(4), 550–66.

Lohmann, L. 2010. Neoliberalism and the calculable world: The rise of carbon trading, in *The Rise and Fall of Neo-Liberalism*, edited by K. Birch and V. Mykhnenko. London: Zed Books, 77–93.

Maasen, S. and Sutter, B. 2007. (Eds.) *On Willing Selves. Neoliberal Politics vis-à-vis the Neuroscientific Challenge*. London: Palgrave MacMillan.

Marzocca, O. 2007. *Perché il governo. Il laboratorio etico-politico di Foucault*. Roma: Manifestolibri.

MacKenzie, D. 2009. Making things the same: Gases, emission rights and the politics of carbon markets. *Accounting, Organizations and Society*, 34(3–4), 440–55.

McCarthy, J. and Prudham, S. 2004. Neoliberal nature and the nature of neoliberalism. *Geoforum*, 35(3), 275–83.

McCormick, S. 2007. Democratizing science movements: A new framework for mobilization and contestation. *Social Studies of Science*, 37(4), 609–23.

Mirowski, P. and Plehwe, D. 2009. (Eds.) *The Road From Mont Pelerin: The Making of the Neoliberal Thought Collective*. Cambridge, MA: Harvard University Press.

Mol, A.P.J. and Jänicke, M. 2009. The origins and theoretical foundations of ecological modernisation theory, in *The Ecological Modernisation Reader*, edited by A.P.J. Mol, D. Sonnenfeld and G. Spaargaren. Abingdon: Routledge, 17–27.

Mol, A.P.J., Sonnenfeld, D. and Spaargaren, G. 2009. (Eds.) *The Ecological Modernisation Reader*. Abingdon: Routledge.

Moore, K., Kleinman, D., Hess, D. and Frickel, S. 2011. Science and neoliberal globalization: A political sociological approach. *Theory and Society*, 40(5), 505–32.

Myers, E. 2008. Resisting Foucauldian ethics: Associative politics and the limits of the care of the self. *Contemporary Political Theory*, 7, 125–46.

Nealon, J. 2008. *Foucault Beyond Foucault. Power and its Intensification since 1984*. Stanford: Stanford University Press.

Nordmann, A. 2004. (Ed.) *Converging Technologies – Shaping the Future of European Societies*. High Level Expert Group 'Foresighting the New Technology Wave'. Luxembourg: Office for Official Publications of the European Communities. Available at: http://ec.europa.eu/research/social-sciences/pdf/ntw-report-alfred-nordmann_en.pdf [Accessed: 20 December 2011].

Nowotny, H., Scott, P. and Gibbon, M. 2001. *Re-Thinking Science*. Cambridge: Polity Press.

OECD. 2008. *Growing Unequal? Income Distribution and Poverty in OECD Countries*. Paris: Organisation for Economic Cooperation and Development.

OECD. 2009. *The Bioeconomy to 2030. Designing a Policy Agenda. Main Findings and Policy Conclusions*. Paris: Organisation for Economic Cooperation and Development.

Ong, A. 2006. *Neoliberalism as Exception*. Durham, NC: Duke University Press.

Ottinger, G. 2010. Buckets of resistance: Standards and the effectiveness of citizen science. *Science, Technology, & Human Values*, 35(2), 244–70.

Parry, S. and Dupré, J. 2010. (Eds.) *Nature after the Genome*. London: Wiley.

Peck, J. 2010. *Constructions of Neoliberal Reason*. Oxford: Oxford University Press.

Peck, J., Theodore, N. and Brenner, N. 2009. Postneoliberalism and its malcontents. *Antipode*, 41, 94–116.

Peck, J. and Tickell, A. 2002. Neoliberalizing space. *Antipode*, 34(3), 380–404.

Peet, R. 2007. *Geography of Power: The Making of Global Economic Policy*. London: Zed Books.

Pellizzoni, L. 2011a. Governing through disorder: Neoliberal environmental governance and social theory. *Global Environmental Change*, 21(3), 795–803.

Pellizzoni, L. 2011b. The politics of facts: Local environmental conflicts and expertise. *Environmental Politics*, 20(6), 765–85.

Pestre, D. 2009. Understanding the forms of government in today's liberal and democratic societies: An introduction. *Minerva*, 47, 243–60.

Plehwe, D., Walpen, B. and Neunhöffer, G. 2005. (Eds.) *Neoliberal Hegemony. A Global Critique*. London: Routledge.

Polanyi, K. 1944. *The Great Transformation*. Boston: Beacon Press.

Pollock, N. and Williams, R. 2010. The business of expectations: How promissory organizations shape technology and innovation. *Social Studies of Science*, 40(4), 525–48.

Rajan, K.S. 2006. *Biocapital. The Constitution of Postgenomic Life*. Durham, NC: Duke University Press.

Raman, S. and Tutton, R. 2010. Life, science, and biopower. *Science Technology & Human Values* 35(5), 711–34.

Roco, M. and Bainbridge, W. 2002. (Eds.) *Converging Technologies for Improving Human Performance*. Arlington, VI: National Science Foundation.

Available at: http://www.wtec.org/ConvergingTechnologies/1/NBIC_report.pdf [accessed: 20 December 2011].
Rose, N. 1996a. Governing 'advanced' liberal democracies, in *Foucault and Political Reason*, edited by A. Barry, T. Osborne and N. Rose. London: UCL Press, 37–64.
Rose, N. 1996b. *Inventing Ourselves*. Cambridge: Cambridge University Press.
Rose, N. 2007. *The Politics of Life Itself*. Princeton, NJ: Princeton University Press.
Rouvroy, A. 2008. *Human Genes and Neoliberal Governance. A Foucauldian Critique*. New York: Routledge.
Rutherford, S. 2007. Green governmentality: Insights and opportunities in the study of nature's rule. *Progress in Human Geography*, 31(3), 291–307.
Saad-Filho, A. and Johnston, D. 2005. (Eds.) *Neoliberalism. A Critical Reader*. London: Pluto
Steger, M. and Roy, R. 2010. *Neoliberalism. A Very Short Introduction*. Oxford: Oxford University Press.
Suarez-Villa, L. 2009. *Technocapitalism*. Philadelphia: Temple University Press.
Swyngedouw, E., 2010. Apocalypse forever? Post-political populism and the spectre of climate change. *Theory, Culture & Society*, 27(2–3), 213–32.
Szerszynski, B. and Urry, J. 2010. (Eds.) Changing climates. *Theory Culture and Society*, 27(2–3), special issue.
Tazzioli, M. 2011. *Politiche della verità. Michel Foucault e il neoliberalismo*. Verona: Ombre Corte.
Tickell, A. and Peck, J. 2003. Making global rules: Globalization or neoliberalization?, in *Remaking the Global Economy*, edited by J. Peck and H.W. Yeung. London: Sage, 163–81.
Tyfield, D. 2010. Neoliberalism, intellectual property and the global knowledge economy, in *The Rise and Fall of Neo-Liberalism*, edited by K. Birch and V. Mykhnenko. London: Zed Books, 60–76.
Venn, C. and Terranova, T. 2009. (Eds.) Thinking after Michel Foucault. *Theory Culture & Society*, 26(6), special issue.
Virno, P. 2004. *A Grammar of the Multitude. For an Analysis of Contemporary Forms of Life*. Los Angeles, CA: Semiotext(e).
Virno, P., 2009. Natural-historical diagrams: The 'new global' movement and the biological invariant. *Cosmos and History*, 5(1), 92–104.
Ward, K. and England, K. 2007. Introduction: Reading neoliberalization, in *Neoliberalization: States, Networks, Peoples*, edited by K. England and K. Ward. Oxford: Blackwell, 1–22.
Williamson, J. 1993. Democracy and the 'Washington consensus'. *World Development*, 21(8), 1329–36.

PART 1
Neoliberalism, technoscience and late capitalism

Chapter 1
Neoliberalism and technology: Perpetual innovation or perpetual crisis?

Laurence Reynolds and Bronislaw Szerszynski

Introduction

The neoliberal era has often been imagined as a period of intense technological revolution. The goal of a high-tech 'knowledge-based economy' (KBE) of perpetual innovation has been elevated into a key guiding principle and salvationary strategy for advanced capitalist economies. Innovation is held up as the solution to multiple problems that became apparent in the 1970s, including the crisis in capital accumulation, the globalization of competition and the rise of environmental degradation. Since that decade we have been dazzled by a seemingly escalating proliferation of innovations, from information technology and mobile telephony through to biotechnology and nanotechnology. Yet, at the same time, there is also a sense that the high-tech promise of the 1970s, of a 'space age' where robots would replace workers, or the later prediction of a 'biotech century', have somehow not been realized. 'Tomorrow's world' never quite came about.[1]

Sceptical commentators have questioned the idea of the technological fecundity of the neoliberal period, arguing instead that the last quarter of the twentieth century and after has been a 'great stagnation', where we have reached a 'technological plateau' (Cowen 2011). A popular trope, deployed by both Gordon (2000: 60) and Cowen (2011), has been to compare the dramatic changes in everyday life that resulted from technological transformations during the first half of the twentieth century with the much more modest changes experienced since then. Save for the Internet and mobile telephony, the basic technological infrastructure (based on cars, oil, etc.) has seen little radical transformation. In this trope, our contemporary experience is indeed one that looks like a plateau when compared with the radical techno-social change that someone reaching old age in the 1960s would have experienced over the preceding half century.

We will explore this apparent paradox in this chapter, with our main focus not on the innovation of new consumer goods, but on what Marx called 'the forces of production'. It is useful here to introduce Marx's own distinction between

1 *Tomorrow's World* was a futurological UK BBC television programme on the impact of new technology that was first aired in 1965, reached its heyday in the 1970s, and ran until 2003.

goods produced for 'Department One' and those for 'Department Two', where the former denotes products that are means of production, and the latter consumer products. Our task therefore is to interrogate whether the period emerging from the crisis years of the early 1970s that has come to be called 'neoliberalism' saw the assemblage of a new set of productive forces. To what extent was the apparent success and renewed global expansion of capitalism in the neoliberal period based upon a new 'techno-economic paradigm' or 'third industrial revolution'?

As we shall see, for sceptical commentators such as Smil (2005) and Gordon (2000), what was heralded in the early 1990s as a 'new economy' based upon infotech and biotech is in no way comparable with the substantial development of the forces of production that was achieved in what has been dubbed the 'second industrial revolution', which began in the late nineteenth century and reached full fruition in the decades following the Second World War. Optimistic talk in the 1970s of a 'third technological revolution' is, in this view, misplaced.

We begin by locating the neoliberal period within an analysis of the different terms that scholars have used to describe the systemic shift in the structure of capitalism in the 1970s. We then go on to critically examine the claims that this shift was made possible by a 'third technological revolution'. In the 1990s, capitalism appeared to recover from its two decades of stagnation and went on to have two decades of growth, a revival which was purported to constitute a fifth Kondratieff wave made possible by a new suite of technologies. We bring sceptical commentators to bear on this account, examining the weak performance of the high-technology sector in terms of productivity. We also survey some of the factors that may lie behind this – and also behind the resumption of growth that occurred nevertheless. We then set the transformed relationship between science and capitalism in the neoliberal period in historical context, by examining their couplings in earlier techno-economic regimes. We conclude by using the story of an 'innovation plateau' to describe the two faces of the relationship between science and capitalism in the neoliberal period: on the one hand, an economy largely characterized by mundane technologies and globalization, and on the other a scientific commons continually appropriated and harvested by capital and caught up in political economies of promise.

The crisis of the seventies and the end of the 'Golden Age'

'Neoliberalism' is one way of naming a set of strategic responses by states, corporations and other actors to the crisis decade of the 1970s (Harvey 2005). The term usually denotes the eventual new direction taken after the crisis and collapse of the Keynesianism or organized capitalism during the global political turbulence opened by the insurgent year of 1968, the collapse of the Bretton Woods international order and the OPEC 'Oil Shock'. The year 1973 is frequently taken as marking a key turning point (e.g. Harvey 1989, Jameson 1991), dividing the second half of the twentieth century into two periods. The first, described variously as 'the long boom' or the 'Golden Age' of capitalist expansion, roughly falls between the end of the Second World War and the 1970s, (Marglin and Schor 1992, Brenner 2002a). The second period, from around 1973 to the present moment, has been characterized as one dominated by 'neoliberalism', a period of economic and political restructuring, privatization, deregulation and globalization in response to the 1970s crisis.

However the question of precisely how to characterize these two periods has brought forth a range of theorizations and terminologies. Some of these focus on state–economy relations, others highlight the techno-economic, others finance, politics or culture. For Lash and Urry (1987), the post-war 'Golden Age' period marked the zenith of 'organized capitalism', a term derived from Hilferding ([1910] 1981) and Bukharin ([1918] 1972) to describe the tendency from the turn of the twentieth century onwards towards the concentration of capital, leading to the fusion of industrial, banking and commercial capital with the state. Here, competition between firms becomes transformed into imperialist competition between states, and the state increasingly steps in to organize the economy and society, creating scientific and bureaucratic hierarchies to manage industrial production and social reproduction. This organization of capitalism, while initially performed by the state and managerial elites, is progressively also taken up by organized labour and civil society.

In a similar vein, for Ruggie (1982) and Harvey (2005) the pre-1970s 'Golden Age' can be described as one of 'embedded liberalism'. Ruggie uses this term to describe the international order established in response to the mid-twentieth-century catastrophes of global depression and war, leading to a new post-war order that was neither disembedded market liberalism nor a protectionist and autarkic state-led monopoly capitalism. Other descriptions name this as a Keynesian or Welfare State period, while others still describe the importance of this *dirigiste*, state-led economic form in developing countries between 1945 and 1970 (Kennedy 2006). According to all of these theorizations, however, by the 1970s a period of change had set in.

Similarly, the naming of the period since the 1970s as 'neoliberalism' is only one attempt at description amongst many others. Thus it has been proposed that the period also marks the beginning of a 'post-industrial society' or an 'information age' (Drucker 1969, Toffler 1970, Bell 1973, Castells 2000), 'postmodernity' (Lyotard 1984, Baudrillard 1984, Jencks 1986, Harvey 1989, Jameson 1991) or 'post-Fordism' (Piore and Sabel 1984, Aglietta 1979, Lipietz 1982, Lash and Urry 1987, Jessop 1992). More recently, many of these theories have been combined in the policy discourse of a 'knowledge-based economy' (OECD 1996), a concept which bears an uncanny resemblance to autonomist Marxist accounts of a new period of capitalism based upon the exploitation of a 'general intellect' (Hardt and Negri 2000, Tronti 1966, Virno 2001) or 'cognitive capitalism' (Moulier-Boutang 2012, see also Gandini this volume). Theorists of social change have often been primed by a cultural expectation that we should be living through a period of 'epochal' changes, where these expectations are combined with the explanation of them. These works therefore inevitably move between descriptive and performative dimensions, being inescapably bound up with the changes themselves and becoming crucial in the self-understanding of the actors involved in these transformations. Hardt and Negri declare this period marks a 'qualitative passage in modern history' from the standpoint of historical materialism, adding that:

> When we are incapable of expressing adequately the enormous importance of this passage, we sometimes quite poorly define what is happening as the entry into postmodernity. We recognize the poverty of this description, but we sometimes prefer it to others because at least postmodernity indicates the epochal shift in contemporary history. (Hardt and Negri 2000: 237)

Later, they add that 'postmodernization' is synonymous with 'the informationalization of production' (Hardt and Negri 2000: 280). For us this raises the question of the relationship between science, technology and this alleged epochal shift into a new phase of capitalism that began around the crisis decade of the 1970s.

In other readings, this decade of crisis takes on significance as a moment of transition between two 'long waves' of economic development. Long wave theory originates with Nikolai Kondratieff (1925), who noted decades-long cycles in prices, interest rates, trade and production. With an average length of around 50–60 years, each 'Kondratieff wave', or K-Wave, is seen as made up of two main phases, an 'upswing' or 'expansionary' phase, and a 'downswing' or 'depressive' phase (Mandel 1995: 20–21). For some long-wave theorists, such as Joseph Schumpeter (1934), Christopher Freeman (Freeman and Louçã 2001) and Carlotta Perez (2002), these cycles are driven by technological innovation. In this view, each upswing phase of a K-Wave requires a new industrial revolution, one in which a 'lead industry' (Rostow 1978) or 'general-purpose technology' (Bresnahan and Trajtenberg 1995) generalizes a whole new set of technological productive forces across the economy. Technologies that have been claimed to have performed that kind of catalytic role, such as steam power, railways and electric motors, are, according to most commentators, essential for an industrial revolution.[2]

Such understandings of Kondratieff waves are highly relevant to our understanding of the neoliberal period. According to proponents of K-Waves, in the 25 or so years after the Second World War the economy was undergoing the expansionary upswing of the fourth K-wave, based on oil, mass production and petrochemicals, and the crisis of the 1970s represents the start of its depressive downswing. But, for neo-Schumpetarians such as Freeman and Perez, this crisis decade also marks the start of a fifth wave, one based around a new

2 Other commentators, however, are sceptical about such 'innovation-centric' accounts, arguing that if a purported 'general purpose technology' had not been available, in most cases its function could have been filled by an alternative, existing one (Fogel 1964, Fishlow 1965, Edgerton 2006).

suite of technologies such as information and communication technologies and biotechnology. It is this kind of claim that we are interrogating here. Are there any signs of new lead industries or general-purpose technologies which might carry a new long wave of economic expansion? Or will economic growth and increased productivity have to come from within the existing techno-economic paradigm? To fully answer – or even articulate – such questions we have to attend to the difficult question of *why* expansive waves start and end.

For Mandel (1995), expansive periods enter periods of crisis and a downswing because of endogenous factors generated by the internal contradictions of capitalist development, such as the tendency of the rate of profit to fall. However, the transition to a new expansive phase is the outcome of a cluster of historically contingent or exogenous factors. These include wars, revolutions, and crises where capital is destroyed and/or the costs of labour are massively reduced. Other factors behind the start of a new K-wave can include the innovation of a new assemblage of productive forces or techno-economic relations. Mandel's analysis helps us hone our question. If capitalist development went into an era of crisis in the 1970s (due to the exhaustion of the previous long wave), then what accounts for the much heralded period of expansion in the 1990s and after? In particular, was there a new assemblage of techno-economic relations sufficient to restore the rate of profit?

A third industrial revolution?

The moment of crisis and transition in the 1970s was understood by many participants and commentators at the time as a new technological revolution. A cluster of new technologies around the microprocessor opened the prospect of a fundamental reordering of economic and social relationships that was billed as the coming of an 'information society' (Bell 1973). In this vision, 'smart production' would see workers replaced by robots, while the informationalization of the forces of production would make them both more resource-efficient and flexible. Following from the 1960s rhetoric of an economy re-forged in the 'white heat' of technological revolution, nuclear power and the space age, expectations were high. In time, the nascent sectors of biotechnology and nanotechnology would be placed within this growing pantheon of promised high-tech futures, a pantheon that seemed to offer salvation not only from the economic quagmire but from the environmental crisis, recognition of which began to proliferate at this time. For some, this heralded a coming 'third industrial revolution', as significant as the second industrial revolution that began 100 years earlier in the 1870s and had profoundly shaped the twentieth century. Was the period of neoliberal globalization and economic growth after the 1970s based upon this promised high-tech 'third industrial revolution' – or were some other factors at play? In this section we discuss this question, focusing on the cases of ICT and biotechnology. Did either of these have the potential to become a 'lead industry' or 'general-purpose technology', in the way that has been claimed about steam power or the electric motor in earlier periods of strong economic growth?

The claims of ICT to be a general-purpose technology has had to contend with what Robert Solow (1987) called the 'productivity paradox'. Solow pointed out that the time of proliferation of computers in the 1970s and 1980s was also a time of slowing productivity in the USA and more widely. Solow's paradox has been much debated since. David (1990) countered by arguing that general-purpose technologies always have a delay before their productivity benefits are realized, citing the low productivity growth in the United States between 1900 and 1920 despite the existence of electric dynamos and motors. The productivity gains from IT did seem to start to be realized in the mid-1990s, when the USA experienced a productivity surge that some analysts claimed was entirely due to the application of IT (Rhode and Toniolo 2006: 10–11). Yet others argue that the productivity gains associated with computers have been confined to the IT sector itself and to computer-intensive parts of the economy (Gordon 2000, Field 2006). Overall, multi-factoral productivity (MFP) growth in the 1990s was in fact lower than that in the 1930s; and in the 88% of the economy that lies outside durable manufacturing, MFP growth *decelerated*, despite investment in computers (Gordon 2000: 72).

Rather as Fogel and Fishlow argued about the role of railways in the nineteenth century, Field suggests that much of the productivity gains in the wider economy could well have happened without computers; many of those productivity gains are being driven not by technological change itself but by changes in unit size such as big-box retailing (Field 2006: 109). Gordon (2000) provocatively suggests that that the steadily declining cost of computing power, far from indicating the strength of computers as a factor of production, in fact point to its weakness. The great inventions of the second industrial revolution did not experience the same continuous price decline, because the significant productivity gains provided by their adoption meant that demand kept up with supply. Gordon uses an analysis of the relationship between supply and demand in computing from 1963 to 1999 to argue that the main productivity gains offered by computers were realized early on their diffusion. For example, the productivity gain from developments in word processing technologies flattens out quickly after the development of automatic reformatting and cut-and-paste. As he argues, '[t]he fixed supply of time to any individual creates a fundamental limitation on the ability of exponential growth in computer speed and memory to create commensurate increases in output and productivity' (Gordon 2000: 62).

Biotechnology offers a similar story of widespread excitement and hype but little real impact on productivity. The turn of the twentieth century saw popular commentaries proclaiming *The Biotech Century* (Rifkin 1998) or *The Biotech Age* (Oliver 2003) – the former reacting with horror and the latter with hope to the prospect. In the early years of the present century, the discourse of the 'knowledge-based bioeconomy' (KBBE) was adopted by transnational policy organizations and the biotechnology industry. As the European bio-industry association puts it:

> In the 18th and 19th Centuries, European society was transformed by the Industrial Revolution and the steam engine. This was the Age of Engineering. In the 20th Century, the developed world reaped the benefits of chemistry, which provided the materials, productive agriculture and medicines which make our lives so comfortable and safe. The whole world is now in transition from the Age of Chemistry to the Age of Biotechnology. (EuropaBio 2005)

The discourse of the knowledge-based bioeconomy was embraced by the OECD, who defined the bioeconomy as 'the aggregate set of economic operations in a society that use the latent value incumbent in biological products and processes to capture new growth and welfare benefits for citizens and nations' (OECD 2005: 3), and by the European Union, who insisted that the KBBE would 'lead to the creation of new and innovative goods and services that will enhance Europe's competitiveness and meet the needs of its citizens' (DG Research 2005a: 3).

Yet despite massive investment of public and private money, the biotechnology revolution has not resulted in increased productivity or profitable commodity production. Wallace (2010) reports that only two medical biotech companies (Amgen and Genentech) and one agricultural biotech company (Monsanto) have made significant profits from selling commodities. Apart from the biggest biotech firm, Amgen, the medical biotech industry has made steady losses throughout its history. Although the 'biotech revolution' has accelerated drug discovery, this has not followed through into drug development and clinical practice, so has failed to reverse the decline in productivity of the pharmaceutical sector (Nightingale and Martin 2004). Despite heavy investment, profitability in the industry has been flat for over thirty years. As far as agricultural biotech is concerned, the industry has produced only two widely used traits – herbicide tolerance and insect resistance – and the evidence of increased agricultural productivity thanks to GM crops engineered to have these traits is ambiguous at best (Wallace 2010: 115, 119–21).

Thus, despite the proliferation of consumer electronics, the contemporary new knowledge economy has so far not produced anything equivalent to the paradigm-shifting technologies of earlier industrial revolutions. It is still possible that new, paradigm-shifting innovations will emerge in the next few decades from research in areas such as materials science and energy. But as mentioned above, in many ways, contemporary life in general and economic activity in particular is still fundamentally shaped by the cluster of hugely significant inventions that emerged in the second industrial revolution of the late nineteenth and early twentieth century: electrical power and motors, organic chemistry and synthetics, and the internal combustion engine (Landes 2003). The years between 1864 and 1917 constituted an 'age of synergy' in which the technological foundations for the twentieth century were laid (Smil 2005, 2006), making possible the 'long boom' from the 1940s to the 1970s, with its massive improvements in productivity, health, and standard of living. Yet while we are still living fundamentally within this same world, the capacity for it to continue to produce a stream of significant innovations seems to be becoming exhausted.

Technological innovation has been held up as the solution to economic crisis. In past such crises we have seen the emergence of new suites of technologies, which have had the capacity to restart faltering productivity growth and restore profit rates. However, as we have seen above, it is by no means clear that the neoliberal era managed to find such technologies after the crisis of organized capitalism in the 1970s. There have been a number of candidates for such technologies, but in each case there is as yet little sign of them making a real contribution to increased productivity. Nevertheless, capitalism did experience a new period of expansion globally after its crisis of the 1970s. How did this happen?

There are many factors underlying the apparent expansive period of capitalist development that followed the crisis of the 1970s. In this period, capital was unable to confidently invest in the development of the productive forces in the old industrial centres of the West, given the already high organic composition of capital (the ratio of non-labour costs to labour costs). Instead, capital sought other ways of reproducing itself, beyond investing in new cycles of commodity production within new techno-social labour processes. One of these strategies did indeed involve world-transforming technologies – however, these were not the high-technology productive forces imagined by 1970s futurologists. Instead, they were far more mundane technologies, whose roots lay in far earlier ages of technological innovation. Levinson (2006) draws our attention to shipping containerization, the distribution system based upon the standardized and ubiquitous rectangular metal box that could fit on lorries, ships and trains, thus facilitating a massive acceleration of the global flows or mobilities of matter and commodities. Likewise Smil (2010) directs our gaze to the diesel engines and gas turbines that propelled the container ships and air traffic as the 'prime movers of globalization'. These technologies allowed the global respatialization of the existing productive forces associated with the Fordist era, creating a new, respatialized, globalized Fordism. Thus instead of the 1970s visions of 'tomorrow's world' with fully automated production or bio-factories, the more prosaic reality was one where low paid workers in China and other parts of South East Asia worked with old, second-industrial-revolution productive forces and labour processes, made more productive via a global respatialization. This move, as part of a new 'spatio-temporal fix' (Jessop 2006, see also Harvey 1982) could alter the organic composition of capital by massively reducing labour costs, and facilitate the expansion of the world economy from the 1990s onwards. Of course, this global respatialization of 'second industrial revolution' techno-economic processes that gathered pace from the 1980s onwards was partly enabled by information technology. The material global flows of shipping containers, each unit travelling from one specialized location to another, required logistical governance that was made more feasible through information technology – but not impossible without it. Our claim here is that fundamental to the period of renewed capitalist growth around the closing decade of the twentieth century was a relatively 'low-tech' ensemble of cranes, diesel engines, containers, ships and logistical information technologies, not the spawn of the 'high-tech' R&D labs of the 'knowledge-based economy'.

The post-1970s period also saw the phenomenon of 'financialization', where greater returns on capital could be found through global financial innovations such as futures and derivatives than in the production and exchange of material commodities. Brenner (2002a) argues that from the 1990s the continuation of capital accumulation on a world scale was dependant on a historic wave of speculation and a succession of asset price bubbles, first around the stock market, then housing and credit, noting the role of the US Federal Reserve and various state agencies in nurturing those bubbles. In this form of accumulation, instead of money (M) being invested in the production of commodities (C) to produce profits (M') according to Marx's formula of M-C-M', money was used to generate a profit directly, through interest or financial speculation (M-M'). Technoscience – and the dreams, hopes, promises and expectations that revolve around it – were also caught up in these waves of speculative bubbles, most famously with the launch of Nasdaq and the dot.com bubble, but also the 'genomics bubble' that followed the sequencing of the human genome (Gisler, Sornette and Woodard 2010).

Thus, despite its failure to develop radical new productive forces, technoscience nevertheless has offered a number of mechanisms whereby profit rates could be restored, even if only temporarily. As well as the speculative bubbles discussed above, technological innovation has also created opportunities for higher-than-average returns on investment through corporate concentration, intellectual property and 'technological rents', but these have generally operated by capturing value from elsewhere in the global economy rather than creating new surplus value. But why has capitalism seemed unable to develop a new suite of technologies capable of recreating the strong economic growth of the post-war decades? There is not space in the current chapter to explore this in any depth, but there are clearly a number of interacting factors – some of which are internal to the logic of capital accumulation or technological innovation, while others are external, contingent factors.

For example, Tyler Cowen (2011) argues that by the 1970s developed economies such as the USA had already picked all the 'low-hanging fruit' of economic growth – available land, scientific and technological breakthroughs, and universal education – and that it was all but inevitable that growth would falter. Brenner (2002b: 9–10) uses studies by Zvi Griliches and others to argue there was no decline in the pace of scientific invention after 1973, just a decline in the innovation and diffusion of new technologies. He suggests that persistent over-capacity in global manufacturing industries had led to low profits, and thus to low levels of investment and the stalling of processes of technological change (Brenner 2002a). Webber and Rigby (1996: 492–3) argue that the 1970s was a victim of Marx's 'law of the tendency of the rate of profit to fall' (Marx 1981), which describes how investment in new technologies, while increasing output per worker, has the effect of increasing the organic composition of capital, putting a downward pressure on surplus value and thus profit rates. They argue that the fall in profit rates in advanced economies from the 1970s onwards was because 'increases in the amount of plant, equipment and raw materials per worker were not offset by improvements in efficiency', thus starving industry of investment in new technologies. Arthur (1994) invokes more general laws of technological development, identifying the way that scale economies, learning effects, adaptive expectations and network effects all work to decrease the incentives for individual economic actors to adopt new technologies, and thus to 'lock in' incumbent technologies.

In reality, it is likely that all of these factors, and the interactions between them, have played a role in the production of the late twentieth-century technological plateau in terms of productive forces. But it is also important to attend in more detail to the changing relations between science and capitalism in the neoliberal period, which process interacted in complex ways with the dynamics that have just been described – sometimes clearly a response to the failure of science to produce new 'general-purpose technologies' for the economy, and sometimes perhaps contributing to that failure. In the next section we thus turn to these questions, and try to put these changing relations in a longer historical context.

Science and neoliberalism

As we suggested in the previous section, in the period since the 1970s science, too, became embroiled in processes of financialization, being drawn into new 'political economies of promise'. This involved attracting venture capital, corporate and public funds for speculative new technoscientific developments. Rather than science being part of M-C-M', as a source of new processes and products, it became more caught up in M-M', as an object of speculation. Thus instead of being simply involved in new rounds of general accumulation via the innovation, production and sale of new commodities, science also became more deeply embroiled in new rounds of speculative appropriation. In the neoliberal 'knowledge-based economy', as well as being a means of production generating new products and processes, science becomes a product – and a commodity – in itself.

But for something as intangible and collective as scientific knowledge to become a product or commodity, it must be enclosed, stabilized and held in place by a legal infrastructure of patents, intellectual property laws and other institutional arrangements (see also Pellizzoni and Ylönen this volume). Starting in the 1980s, scientific research, especially in universities and other public institutions began a process of reorganization to fit it ever more closely to the needs of industry and tropes of global competitiveness (Etzkowitz, Webster and Healey 1998, Kenney 1986, Kleinman and Vallas 2001, Kleinman 2003, Slaughter and Rhoades 2004). Thus the landmark Supreme Court decision of *Diamond v. Chakrabarty* in 1980 enabled the patenting of life, and was part of a massive extension of the appropriation of science as intellectual property, while in the same year universities were made to become owners of intellectual property with the 1980 Bayh-Dole Act. A 'university-industrial complex' (Kenney 1986) was consolidated, involving novel couplings and combinations of 'public' and 'private'. The scientific spaces of universities became increasingly like those of corporations – and vice versa – in a process of 'asymmetrical convergence' (Kleinman and Vallas 2001, Vallas and Kleinman 2008). Not only are there more linkages between public and private, between universities and corporations; there are also more linkages or networks between private bodies, such as small scale research 'start-ups' and corporations. All this is to facilitate the flow of knowledge between diverse public and private scientific spaces, between the spaces of its collective and social generation and the spaces of its private appropriation and enclosure.

This transformation of knowledge generation in the neoliberal period may be described as the privatization of science. However, this is a contradictory process, with limits. Knowledge generation is the accumulated effect of networks of cooperation, collective effort and public institutions. It grows through public flows, being a 'non-rival good' whereby multiple users do not consume it but actually generate more knowledge. Merton (1973) recognized the essentially collective and social character of the production of scientific knowledge when he described the norm of 'communalism' in the 'ownership' of scientific discoveries. This leads Jessop (2000) to argue that if in the 'knowledge-based economy' knowledge is a productive force, then it exhibits a tendency (noted by Marx in his time) for capitalism to rely upon increasingly collective and interlinked productive forces. Furthermore, in this Marxian schema these ever more collaborative productive forces stand in ever greater contradiction to their private appropriation. This is a contradiction at the heart of the knowledge-based economy that must be managed. As Jessop notes:

> Knowledge is a collectively generated resource and, even where specific forms and types of intellectual property are produced in capitalist conditions for profit, this depends on a far wider intellectual commons. (Jessop 2002: 129)

This 'wider intellectual commons' is the central resource upon which the appropriation and capital-accumulation strategies of the businesses of the new knowledge economy depend. In their analysis of the role of public science in the creation of biotechnology, McMillan, Narin and Deeds (2000) found that of scientific papers cited in biotechnology patents only 16.5% originated in the private sector, but more than 70% were still the product of scientists working in solely public scientific institutions such as universities. This crucial role of public institutions in generating the knowledge base that private appropriation depends upon underlines the continuing role of the state and public funding. Indeed, the moves mentioned above, such as the reorganization of university-commercial relationships and the legal frameworks supporting intellectual property, formed part of a new state strategy in the neoliberal era. It has often been noted that the neoliberal era did not simply see de-regulation and the 'retreat of the state' but rather complex forms of re-regulation and the reorganization of the public sphere to be optimized for global competition – a form of state intervention named by Cerny (1990) as a 'competition state'. In the field of innovation and science policy, this change has been described as the rise of a 'Schumpeterian Competition State' by Jessop (2002: 96) 'because of its concern with technological change, innovation and enterprise'. In this form, the state attempts to manage the contradictions of the KBE – including those around the commodification of the knowledge commons:

> The state has roles in both regards: it must promote the commodification of knowledge through its formal transformation from a collective resource (intellectual commons) into intellectual property (for example in the form of patents, copyright and licences) as a basis for revenue generation; but it must also protect the intellectual commons as a basis for competitive advantage for the economy as a whole. (Jessop 2002: 129)

Here the 'intellectual commons' – including science – have to be maintained in part as a commons. If the act of appropriating or enclosing encroaches too far into the scientific commons, then they cease to be productive for capital. This can be seen in the problem of 'patent thickets', defined by Carl Shapiro as 'a dense web of overlapping intellectual property rights that a company must hack its way through in order to actually commercialize new technology' (Shapiro 2001: 120).

Thus, in the neoliberal era, there are limits to the full subsumption of science under the logic of capital and its private appropriation as a commodity. This construction of science as a 'commons' points to a kinship between its neoliberal appropriation and what Marx described as 'primitive accumulation' or Harvey (2003) calls 'accumulation by dispossession'. For Marx, this form of accumulation preceded full capitalist accumulation (the extraction of surplus value from labour and the production of commodities described under the algebra of M-C-M'). Instead, primitive accumulation draws value into the capitalist system from 'outside', from pre-and extra-capitalist social forms. Subsequent Marxists have seen this as an ongoing process (De Angelis 2007). Here, instead of fresh sources of accumulation being sucked into the system from an 'external' frontier, the existing capitalist infrastructure (science, education, welfare, etc.) become privatized, reorganized and cannibalized. To describe this, Huws (2012) develops a category of 'secondary primitive accumulation' in contrast to the more traditional forms of 'primary' primitive accumulation. Whereas in primary primitive accumulation value is extracted from natural resources or activities carried out outside the money economy, secondary primitive accumulation, by contrast, involves public services such as education and healthcare. These were developed for their use value in the form of services, arising as non-commodified spaces but still within the capitalist economy. Neoliberal strategies involve the privatization of these services, and their transformation from use value to exchange value. Like public services, science developed as an institution within capitalist society. Both of these kinds of institutions serve the logic of capital accumulation but are never totally subsumed or reducible into it. Science remained as a 'commons', yet unlike with 'primary primitive accumulation' this 'scientific commons' did not pre-exist capitalism but is produced within it. Thus any strategy to enclose aspects of this scientific commons represents a secondary primitive accumulation, in a mode similar to the attempts to exploit public services. Yet as we saw, this commons can never be totally enclosed, or the production of scientific knowledge would be stifled. Rather, the commons must be partly maintained as such – in order to be 'harvested' by the private appropriation strategies of business.

Such theorizations of the 'scientific commons' and 'secondary primitive accumulation' can help us understand not only the neoliberal transformation of science, but also help put these into a much longer historical perspective of the shifting relations between science and capitalism. As the Marxist historian E.P. Thompson once quipped: 'The exact nature of the relationship between the bourgeois and the scientific revolutions in England is undecided. But they were clearly a good deal more than just good friends' (Thompson 1965: 334). Thus the neoliberal era did not witness the first coupling between science and capitalism – this happened at least from the seventeenth century onwards. However, while science and capitalism may have emerged together, the science that emerged played little role in the direct development of capitalism's productive forces at this stage (Toulmin 1992); rather, the relationship in this early period points the other way, with technical achievements providing inspiration for more esoteric theorizing (Hessen 1971, Freudenthal and McLaughlin 2009). Science was therefore an autonomous realm within the developing capitalist society – and still characterized by the efforts of the gentleman amateur. At the same time, as a correlative of this observation, in the development of the technologies of the first industrial revolution a far more decisive role was played by practise, trial-and-error, and the reflexive processes of collective 'tinkering' by many hands and brains located throughout the productive process. Indeed the actual role of 'science' in techno-economic change is argued to have often been either more limited, or different to the imaginings of science as a fecund productive force (Edgerton 2006). It is possible to uncover the more 'mundane' innovations that underlay most eras of imagined technoscientific revolution.

However, the second industrial revolution that began in the late nineteenth and early twentieth centuries marked the enrolment of science (or technoscientific practices and knowledge) ever more directly into the process of production. This was exemplified by the twentieth-century chemicals industry, with its mass of chemists engaged in the innovation of new materials, polymers, pharmaceuticals, agrochemicals, etc. Here, corporate R&D labs and universities were essential in the production of new products and production processes (Mowery and Rosenberg 1989). This was the period of organized capitalism, which had an affinity with the scientific organization of production, welfare and reproduction. Science was reorganized to fit more with this Fordist phase – in terms of its scale (big science) and its industrialized division of labour. Yet at the same time it still kept many of the qualities of being a (relatively) autonomous part of this process – observing its own norms and rituals and not yet commodified in the neoliberal mode we will describe shortly.

Therefore in terms of the direct enrolment of scientific labour as a productive force, the early and mid-twentieth century period of the second industrial revolution saw something of a zenith or high water mark. This stands in contrast to what was to follow from the final quarter of the twentieth century onwards. This new post-1973 period (dubbed variously as 'neoliberalism', 'postmodernity', etc.) was initially imagined as an acceleration of the scientization (or scientific intensification) of the production process, where a third technological revolution would lead to an 'information society' and a 'new economy'. Scientific R&D, it was assumed, would play an increasingly direct role in the innovation of new products and processes. However, as we have argued, the neoliberal era did not see the transcendence of the second industrial revolution by a third, or anything resembling the scale and pace of the techno-economic changes associated with our earlier period. Instead, as we have touched upon above, neoliberal science became caught up in a different dynamic of accumulation. While of course science in the neoliberal period did contribute to some new products and processes, this is eclipsed by another phenomenon – that of science becoming a product itself.

Thus we can provisionally divide the history of the relationship between science and capitalism into three phases, ones which can be specified by drawing on and developing Marx's language of subsumption. In the first volume of *Capital*, Marx (1976) distinguished between the *formal* subsumption of labour to capital – for example, in the 'putting out' system, where there is little attempt by capitalists to shape and optimise artisanal labour taking place in people's homes – and its *real* subsumption – such as in a factory setting, in which complex labour is simplified and there is a huge pressure towards increasing productivity. Such language is helpful in thinking about the changing relations between capitalism and science. In the first phase of this relationship described above, that of the scientific revolution and the first industrial revolution, science is a connected but autonomous realm, one which we could say is only *formally* subsumed to capital. The second phase, that ran from the late nineteenth century up to the 1970s, could be said to be characterized by the increasingly *real* subsumption of science, as scientific research is made central to the process of technological change in the economy, and the scientific labour process is broken up and reorganized under organized capitalism – though retaining its Mertonian norms and thus not totally subsumed under the commodity form. Finally, the third phase, the current one of neoliberalism, has been one in which, on the one hand, science and capitalism have been relatively divorced – a technological revolution without science, and a science without new productive forces – but on the other we have seen the *further* subsumption of science, as science itself is brought under the commodity form, but at the same time the scientific commons becomes even more important to capital, as an extra-capitalist commons generated within capitalism that is 'harvested' rather than totally enclosed.[3]

3 Here we thus depart from both Hardt and Negri's talk of the 'real subsumption of society under capital' (2000: 365), and Liodakis's description of the 'universal (tending to a total) subsumption not only of labour but also of science and nature under capital' (2010: 25). On the subsumption of nature see also Pellizzoni and Ylönen this volume.

Conclusion

When what came to be called neoliberalism emerged in the decades of restructuring following the crisis of the 1970s, it was imagined as a new period of technological innovation – a 'new economy' based on an 'information society', even a 'third industrial revolution'. Yet despite this high-tech imaginary, we have uncovered a contrary story – of an 'innovation plateau' where this promised new technological base revolutionizing the means of production has largely failed to materialize. In this counter-narrative, 'second industrial revolution' technologies continue to predominate, with forms of energy and labour processes already familiar by the mid-twentieth century still producing the bulk of the world's tangible commodities. In his early 1970s work *Late Capitalism* Ernest Mandel (1975) traced the tendency of the abolition of living labour from the labour process by new technology and automation. However, for Mandel the completion of this process of automation was impossible, representing the 'absolute inner limit of the capitalist mode of production' as the elimination of living labour would not only increase the 'organic composition of capital' but abolish the source of surplus value altogether. Commenting on this, Morris-Suzuki (1984) argued that this tendency towards automation would push capital to open up new zones of surplus-value generation where living labour would still be required, namely the innovation process itself, where creativity and science would be key, calling this a 'perpetual innovation economy'.

However, the 'inner limit' of capitalism was not in fact reached; neither did its supersession or amelioration via perpetual innovation unfold. Rather than the 1970s utopian (and dystopian) visions of fully automated production where living labour is abolished or marginalized, a new but 'mundane' ensemble of transportation technologies enabled the global respatialization of the second industrial revolution. Living labour in the form of new working classes in South East Asia and China remains at the centre of neoliberal capitalism. At the same time, while this respatialization accounts for much of the growth in trade and global productivity in the neoliberal period, it does not on its own account for restoration of the rate of profit. Instead, the apparent return to health in the 'boom years' of neoliberalism involved financialization, economies of debt and a succession of speculative bubbles. The financial and broader economic crisis post 2008 represents the shattering of this mirage of the robust health of neoliberal capitalism, suggesting an ultimate failure to escape the problems identified in the 1970s.

This story of an innovation plateau therefore helps us contextualize the role of science in the neoliberal period. Rather than accelerating the tendency of the second industrial revolution to enrol science as a new force of production in a division of labour associated with the vertical corporation, science itself becomes a product. Here it is either enclosed as intellectual property (harvested from a 'maintained commons') or made into the base of (unrealized) promise, producing little more tangible than successful initial public offerings (IPOs) on the stock exchange in the form of 'venture science' (Rajan 2006). Thus science has become locked into the immaterial and speculative economies of neoliberalism's bubbles, in a society that appears unable to innovate its way beyond its economic and ecological crisis in any substantial way. We are left with a picture of neoliberalism involving a relatively mundane or 'low-tech' economic 'revolution' (epitomized by the shipping container) which failed to engage science fully as a productive force on the one hand, and a 'high-tech' world of science that becomes cannibalized and privatized as a commodity on the other hand. This suggests the failure of the modernist dream of the mid-twentieth century; instead of a close articulation between the spheres of scientific knowledge production and economic activity, we see a society stalled on an innovation plateau and consuming itself in cannibalistic privatizations while it entertains itself with the simulation of its once promised high-tech future.

References

Aglietta, M. 1979. *A Theory of Capitalist Regulation: The US Experience*. London: New Left Books.

Arthur, W. 1994. *Increasing Returns and Path Dependency in the Economy*. Ann Arbor: University of Michigan Press.

Baudrillard, J. 1984. On Nihilism. *On the Beach*, 6, 38–9.

Bell, D. 1973. *The Coming of Post-Industrial Society: A Venture in Social Forecasting*. New York: Basic Books.

Brenner, R. 2002a. *The Boom and the Bubble: The US in the World Economy*. London: Verso.

Brenner, R. 2002b. The boom, the bubble, and the future: Interview with Robert Brenner. *Challenge*, 45(4), 6–19.

Bresnahan, T. and Trajtenberg, M. 1995. General purpose technologies: 'Engines of Growth'. *Journal of Econometrics*, 65(1), 83–108.

Bukharin, N.I. [1918] 1972. *Imperialism and World Economy*. London: Merlin Press.

Castells, M. 2000. *The Rise of the Network Society*. Oxford: Blackwell.

Cerny, P.G. 1990. *The Changing Architecture of Politics: Structure, Agency, and the Future of the State*. London: Sage.

Cowen, T. 2011. *The Great Stagnation: How America Ate All the Low-Hanging Fruit of Modern History, Got Sick, and Will (Eventually) Feel Better*. New York: Dutton Books.

David, P. 1990. The dynamo and the computer: An historical perspective on the modern productivity paradox. *American Economic Review*, 80(2), 355–61.

De Angelis, M. 2007. *The Beginning of History: Value Struggles and Global Capital*. London: Pluto.

DG Research. 2005. *New Perspectives on the Knowledge-Based Bio-Economy*. Conference report. Brussels: European Commission, Directorate-General for Research & Innovation.

Drucker, P.F. 1969. *The Age of Discontinuity: Guidelines to Our Changing Society*. New York: Harper & Row.

Edgerton, D. 2006. *The Shock of the Old: Technology and Global History Since 1900*. London: Profile Books.

Etzkowitz, H., Webster, A. and Healey, P. 1998. *Capitalizing Knowledge: The Intersection of Industry and Academia*. Albany: State University of New York Press.

EuropaBio. 2005. *About the Bio-Based Economy*. Available at: http://www.bio-economy.net/bioeconomy/about_bioeconomy/index_aboutbioeconomy.html [accessed: 2 January 2012].

Field, A. 2006. Technical change and US economic growth: The interwar period and the 1990s, in *The Global Economy in the 1990s: A Long Run Perspective*, edited by P.W. Rhode and G. Toniolo. Cambridge: Cambridge University Press, 89–117.

Fishlow, A. 1965. *American Railroads and the Transformation of the American Economy*. Cambridge, MA: Harvard University Press.

Fogel, R. 1964. *Railroads and American Economic Growth: Essays in Econometric History*. Baltimore: Johns Hopkins University Press.

Freeman, C. and Louçã, F. 2001. *As Time Goes By: From the Industrial Revolutions to the Information Revolution*. Oxford: Oxford University Press.

Freudenthal, G. and McLaughlin, P. 2009. *The Social and Economic Roots of the Scientific Revolution*. Dordrecht: Springer.

Gisler, M., Sornette, D. and Woodard, R. 2010. *Exuberant Innovation: The Human Genome Project*. Swiss Finance Institute Research Paper No. 10-12. Available at: http://papers.ssrn.com/sol3/papers.cfm?abstract_id=1573682 [accessed: 2 January 2012].

Gordon, R. 2000. Does the 'new economy' measure up to the great inventions of the past? *Journal of Economic Perspectives*, 14(4), 49–74.

Hardt, M. and Negri, A. 2000. *Empire*. Cambridge, MA: Harvard University Press.

Harvey, D. 1982. *The Limits to Capital*, London: Verso.

Harvey, D. 1989. *The Condition of Postmodernity: An Enquiry into the Origins of Cultural Change*, Oxford: Blackwell.

Harvey, D. 2003. *The New Imperialism*, Oxford: Oxford University Press.

Harvey, D. 2005. *A Brief History of Neoliberalism*, Oxford: Oxford University Press.

Hessen, B. 1971. The social and economic roots of Newton's 'Principia', in *Science at the Crossroads*, edited by N.I. Bukharin. London: Frank Cass, 149–212.

Hilferding, R. [1910] 1981. *Finance Capital: A Study of the Latest Phase of Capitalist Development*. London: Routledge & Kegan Paul.

Huws, U. 2012. Crisis as capitalist opportunity: New accumulation through public service commodification. *Socialist Register* [Online], 48.

Jameson, F. 1991. *Postmodernism, or, the Cultural Logic of Late Capitalism*. Durham, NC: Duke University Press.

Jencks, C. 1986. *What is Post-Modernism?* London: Academy Editions.

Jessop, B. 1992. Fordism and post-Fordism: A critical reformulation, in *Pathways to Industrialization and Regional Development*, edited by M. Storper and A. Scott. London: Routledge, 46–69.

Jessop, B. 2000. The state and the contradictions of the knowledge-driven economy, in *Knowledge, Space, Economy*, edited by J.R. Bryson, P.W. Daniels, N. Henry and J. Pollard. London: Routledge, 63–78.

Jessop, B. 2002. *The Future of the Capitalist State*. Cambridge: Polity Press.

Jessop, B. 2006. Spatial fixes, temporal fixes and spatio-temporal fixes, in *David Harvey: A Critical Reader*, edited by N. Castree and D. Gregory. Oxford: Blackwell, 142–66.

Kennedy, D. 2006. The rule of law, development choices, and political common sense, in *The New Law and Economic Development: A Critical Appraisal*, edited by D.M. Trubek and A. Santos. New York: Cambridge University Press, 95–173.

Kenney, M. 1986. *Biotechnology: The University Industrial Complex*. New Haven, CT: Yale University Press.

Kleinman, D. 2003. *Impure Cultures: University Biology and the World of Commerce*. Madison: University of Wisconsin Press.

Kleinman, D. and Vallas, S. 2001. Science, capitalism, and the rise of the 'knowledge worker': The changing structure of knowledge production in the United States. *Theory and Society*, 30(4), 451–92.

Kondratieff, N. 1925. Long business cycles. *Problems of Economic Fluctuations*, 1, 28–79.

Landes, D. 2003. *The Unbound Prometheus: Technological Change and Industrial Development in Western Europe from 1750 to Present*, second edition. Cambridge: Cambridge University Press.

Lash, S. and Urry, J. 1987. *The End of Organized Capitalism*. Cambridge: Polity Press.

Levinson, M. 2006. *The Box: How the Shipping Container Made the World Smaller and the World Economy Bigger*. Princeton, NJ: Princeton University Press.

Liodakis, G. 2010. *Totalitarian Capitalism and Beyond*. Farnham: Ashgate.

Lipietz, A. 1982. Towards global Fordism? *New Left Review*, 132, 33–47.

Lyotard, J-F. 1984. *The Postmodern Condition: A Report on Knowledge*. Minneapolis: University of Minnesota Press.

Mandel, E. 1975. *Late Capitalism*. London: NLB.

Mandel, E. 1995. *Long Waves of Capitalist Development: A Marxist Interpretation*, revised edition. London: Verso.

Marglin, S. and Schor, J. 1992. (Eds.) *The Golden Age of Capitalism: Reinterpreting the Postwar Experience*. Oxford: Clarendon.

Marx, K. 1976. *Capital*, Vol. 1, tr. Ben Fowkes. New York: Vintage.

Marx, K. 1981. *Capital*, Vol. 3, tr. David Fernbach. New York: Vintage.

McMillan, G., Narin, F. and Deeds, D. 2000. An analysis of the critical role of public science in innovation: The case of biotechnology. *Research Policy*, 29, 1–8.

Merton, R. 1973. The normative structure of science, in *The Sociology of Science*. Chicago, IL: University of Chicago Press, 267–78.

Morris-Suzuki, T. 1984. Robots and capitalism. *New Left Review*, 147, 109–21.

Moulier-Boutang, Y. 2012. *Cognitive Capitalism*. Cambridge: Polity.

Mowery, D. and Rosenberg, N. 1989. *Technology and the Pursuit of Economic Growth*. Cambridge: Cambridge University Press.

Nightingale, P. and Martin, P. 2004. The myth of the biotech revolution. *TRENDS in Biotechnology*, 22(11), 564–9.

Oliver, R. 2003. *The Biotech Age: The Business of Biotech and How to Profit from It*. New York: McGraw-Hill.

OECD. 1996. *The Knowledge-Based Economy*. Paris: Organization for Economic Cooperation and Development.

OECD. 2005. *The Bioeconomy to 2030: Designing a Policy Agenda*. Paris: Organization for Economic Cooperation and Development.

Perez, C. 2002. *Technological Revolutions and Financial Capital: The Dynamics of Bubbles and Golden Ages*. Cheltenham: Elgar.

Piore, M. and Sabel, C. 1984. *The Second Industrial Divide: Possibilities for Prosperity*. New York: Basic Books.

Rajan, K.S. 2006. *Biocapital: The Constitution of Postgenomic Life*. Durham, NC: Duke University Press.

Rhode, P. and Toniolo, G. 2006. Understanding the 1990s: A long-run perspective, in *The Global Economy in the 1990s: A Long-Run Perspective*, edited by P.W. Rhode and G. Toniolo. Cambridge: Cambridge University Press, 1–20.

Rifkin, J. 1998. *The Biotech Century: Harnessing the Gene and Remaking the World*. New York: Jeremy P. Tarcher/Putnam.

Rostow, W. 1978. *The World Economy*. Austin: University of Texas Press.

Ruggie, J. 1982. International regimes, transactions, and change: Embedded liberalism in the postwar economic order. *International Organization*, 36(2), 379–415.

Schumpeter, J. 1934. *The Theory of Economic Development*. Cambridge, MA: Harvard University Press.

Shapiro, C. 2001. *Navigating the Patent Thicket: Cross Licenses, Patent Pools, and Standard-Setting, Innovation Policy and the Economy*. Volume 1, edited by A.B. Jaffe, J. Lerner and S. Stern. Cambridge, MA: MIT Press, 119–50.

Slaughter, S. and Rhoades, G. 2004. *Academic Capitalism and the New Economy: Markets, State, and Higher Education*. Baltimore: Johns Hopkins University Press.

Smil V. 2005. *Creating the Twentieth Century: Technical Innovations of 1867–1914 and Their Lasting Impact*. New York: Oxford University Press.

Smil, V. 2006. *Transforming the Twentieth Century: Technical Innovations and Their Consequences*. New York: Oxford University Press.

Smil, V. 2010. *Prime Movers of Globalization: The History and Impact of Diesel Engines and Gas Turbines*. Cambridge, MA: MIT Press.

Solow, R. 1987. We'd better watch out. *New York Times*, Book Review (July 12), 36.

Thompson. E. 1965. The peculiarities of the English. *Socialist Register*, 2, 311–62.

Toffler, A. 1970. *Future Shock*. New York: Random House.

Toulmin, S. 1992. *Cosmopolis: The Hidden Agenda of Modernity*. Chicago, IL: University of Chicago Press.

Tronti, M. 1966. *Operai e Capitale*. Torino: Einaudi.

Vallas, S. and Kleinman, D. 2008. Contradiction, convergence, and the knowledge economy: The co-evolution of academic and commercial biotechnology. *Socio-Economic Review*, 6(2), 283–311.

Virno, P. 2001. General intellect, in *Lessico Postfordista*, edited by A. Zanini and U. Fadini. Milano: Feltrinelli.

Wallace, H. 2010. *Bioscience for Life? Who Decides What Research Is Done in Health and Agriculture?* Buxton: GeneWatch UK.

Webber, M. and Rigby, D. 1996. *The Golden Age Illusion: Rethinking Postwar Capitalism*. New York: Guilford Press.

Chapter 2

Hegemonic contingencies: Neoliberalized technoscience and neorationality

Luigi Pellizzoni and Marja Ylönen

Introduction

Studies of neoliberalism tend to polarize between two different interpretations of it: as an ideology or discourse entailing a particular description of individuals, society, nature and related issues, or as an economic-political process (globalization, the triumph of capitalism over the state, the expansion of economy over traditional politics). At the same time, Foucauldian and neo-Marxist readings of neoliberalism have often been depicted as incompatible (Barnett 2005). However, despite conceptual and methodological discrepancies, we concur with those who believe in the fruitfulness of theoretical cross-fertilizations (e.g. Larner 2003, Ward and England 2007, Jessop 2007). In this chapter we elaborate on the governmentality approach while at the same time benefiting from neo-Marxist insights into hegemony in order to explore a subtle yet profound change in the way social affairs are understood and organized vis-à-vis their biophysical underpinnings; a change that involves technoscience simultaneously as a source and as a core object.

This is not the place to elaborate on the relationship between Foucault and Marx, or on the debates internal to the two scholarships. We instead complement the governmentality framework with dialectical materialist[1] insights into hegemony, in order to capture some aspects of the relationship between neoliberalism and technoscience. Foucault himself used the concept of hegemony.[2] Above all, there is both an affinity and a complementarity between Foucauldian insights into

1 The basic difference between structural Marxism and dialectical materialism lies in the latter's monistic (vs. dualistic) understanding of society. The material and ideal, or economy and culture, belong to the same level of society, being relatively autonomous, though interacting, spheres (Alanen 1991, Sayer 1979). Material relations provide opportunities and limits within which ideas can develop, without determining them, however. All social phenomena have ideal and material aspects, mediated by discourses. Dialectical materialism can thus more easily engage in a dialogue with Foucauldian post-structuralism.

2 For example he refers to 'social hegemony', 'hegemonic effects', 'hegemony of the bourgeoisie' and 'class-domination' (see Foucault 1980: 122, 156, 188; 1982: 224; 2003: 224).

governmental rationality and a dialectical materialist outlook on hegemony. The affinity includes a view of the social order as a continuous struggle and rebalancing of opposing forces and the interweaving therein of material and symbolic elements, the performativity of discourses, and the making sense of practices (Hunt 2004, Lemke 2007). The complementarity is usually found in the Foucauldian emphasis on the 'how' and the Marxist focus on the 'why' of a particular rule; or in the former's focus on discourses and the latter's concern with their connection with material underpinnings (Harvey 1996, Dean 1999, Jessop 2007). Yet we shall see that both frames provide insights into both types of questions, which strengthens the sense of their complementarity.

Our basic argument is that a specific rationality of government underlies the present intertwining, or reciprocal constitution, of neoliberalism and technoscience, and that the features of this rationality significantly contribute to its hegemonic status. We start by outlining the way some elements of hegemony theory can supplement our understanding of neoliberalism and thus complement the governmentality perspective. We then elaborate on a crucial, yet little noticed, implication of such analysis. Neoliberal governmentality exhibits a novel form of rationality: it combines hyper-modernist and post-modernist traits, giving rise to a distinctive way of conceiving and actualizing contingency, instability and agency. Technoscience is profoundly traversed by this logic. Biotechnology regulation and practices offer a paradigmatic example of the neoliberal remoulding of nature, people and politics at the same time. Moreover, 'ethics' and 'dialogue' act as major triggers of the neoliberalization of technoscience. The chapter concludes with a reconsideration of the question of 'resistance' in light of the peculiarly hegemonic character of neoliberal rationality.

Hegemony, governmentality and neoliberalism

Neoliberalism is often portrayed as a project of social change whereby 'human well-being can best be advanced by liberating individual entrepreneurial freedoms and skills within an institutional framework characterized by strong private property rights, free markets, and free trade' (Harvey 2005: 2). From this viewpoint, neoliberalism is first of all an ideology aiming at hegemony (Jessop 2002, Plehwe, Walpen and Neunhöffer 2005, Mirowski and Plehwe 2009).

The Gramscian concept of hegemony refers to ruling not only through coercion but above all through consent. Hegemony is a never-ending process of creation, maintenance and reproduction, challenge and modification of power (Ransome 1992). To obtain power (or domination in Foucault's parlance: see below), a historical bloc[3] needs to develop a universal world-view that compromises with

3 Hegemony develops from an historical bloc which aims to gain leadership in the development of society. There are several historical blocs, yet only one can be the hegemonic bloc, the one provided with economic, political and social leadership. In the

elements of suppressed groups' interests and the hegemonic bloc's interests, without challenging the core economic interests of the hegemonic bloc (Gramsci 1978). Other classes or groups are persuaded that accepting a particular social order makes sense either in factual or in principled terms, or both – it corresponds to reality, it is the best achievable one, it is right and beneficial to everybody. Hegemony, in other words, requires the integration of meanings belonging to different classes and fractions into ethical-political discursive projects, creating 'nodes of interests' (Laclau and Mouffe 1985) on which consent or coercion can be based. Integration legitimizes a biased portrayal of social affairs (Hall 1983, see also Chiapello 2003). Hegemony, however, is more than distorted knowledge, because it is grounded on thought forms (or meaning systems) furnished by myths, religion, popular and – increasingly – scientific narratives, which are objective in their being over-individual and regarded as sufficiently corresponding to reality. Objective thought forms are expressed in discourses, especially those of authoritative people and institutions. Despite their inadequacies and biases as regards relevant knowledge and social relationships, they provide people with a rational interpretive framework tested in and adapted to everyday life (Alanen 1991, Hall 1983, Ylönen 2011). Discourse formation is related to people's positions in society, because these positions incorporate concrete practices which limit and validate the content of discourses. In this sense, neoliberalism as a discourse cannot be detached from capitalist material practices (Robinson 2004, cf. also Harvey 1996: 78–92).

Hegemonic constructions use a variety of available objective thought forms, combining different and even opposed ideas, norms and practices.[4] A first clue to the hegemonic status of neoliberalism, then, is its capacity to evolve and adapt to changing institutional and cultural contexts and conditions (see e.g. Jessop 2002, Ong 2006, Brenner, Peck and Theodore 2010). This chameleonic character hides unity behind variety, which makes neoliberal politics and policies seem to be pragmatic responses to objective yet variable problems of globalization, economic growth or ecological threats.[5]

Hegemony in its cultural aspects – as an ongoing process based on objective thought forms that trigger and support nodes of interests – contributes to Foucault's governmentality perspective. Despite their difference of emphasis on the ideational and praxeological aspects of power, the Foucauldian and the neo-Marxist

case of neoliberalism, transnational actors in business, bureaucracy and intellectual spheres play an especially relevant role (Ransome 1992, Plehwe, Walpen and Neunhöffer 2005, Mirowski and Plehwe 2009).

4 Consider only the contradiction between the neoliberal aspiration to increased individual freedom and reduced state intervention and the current widespread securitarian drift, with a dramatic expansion of surveillance technologies (Mattelart 2010).

5 Paradoxically this variety, as reflected in the variegated assemblages of policies, practices and discourses which compose the landscape of neoliberalization processes (Brenner et al. 2010), makes for some scholar inappropriate to resort to the concept of hegemony (Barnett 2005).

approaches intersect in stressing the role of political rationalities in providing cognitive and normative maps that allow political actors to develop strategies for realizing goals. As the word 'governmentality' implies, this perspective focuses on the rationality of rule, not in terms of a transcendental reason but in those of specific forms of reasoning around which the exercise of power is articulated (Foucault 1991, Dean 1999, see also Bárd this volume, Ferreira et al. this volume). The notion of government, therefore, refers to these reflective modes of power, which make use of science but also of routines, normative orientations and cultural self-evidence (Lemke 2007) – objective thought forms in the parlance of hegemony theory. Government is the regulation of conduct by a more or less rational application of technical means deemed appropriate to the goals pursued (Hindess 1996: 106). It pertains to an economy of conduct: the conduct of the conduct of people, insofar as they are free agents. In this sense, government is the linking element between, on the one hand, power as the strategic games between liberties inbuilt in human relationships, and on the other, states of domination where, through material-symbolic technologies, power relations are crystallized in constantly asymmetrical, difficult to reverse, ways (Foucault 1997a). Government thus fits between consent and coercion (Hindess 1996: 67), both of them becoming instruments, rather than sources, of power (Foucault 2000). The conduct of conduct can be seen as based on nodes of interests between those who govern and those who are governed. Then if hegemony has integrative, legitimizing and distorting aspects (Gramsci 1978, Ransome 1992, Chiapello 2003), it is the latter that is most directly related to the shaping or controlling of the conduct of the governed.

Foucault's analysis of government is also a genealogical reconstruction of the modern state. Governmentalization, then, means the predominance of the 'administrative state' based on professional knowledge and expert systems over the old 'state of justice' based on sovereign power, resulting 'in the formation of a whole series of specific governmental apparatuses and […] in the development of a whole complex of *savoirs*' (Foucault 1991: 103). It means a growing political role of truth, as established by experts and sciences (Foucault 1997b).

Foucault's reading of governmental change, moreover, is conducted in terms of 'intensification'. 'The emergence of new modes of power happens through the lightening, saturation, becoming-more-efficient, and transversal linkage of existing practices […] [up to] tipping points […] where the object or subject mutates into another form' (Nealon 2008: 38–9). Neoliberal governmentality can therefore be interpreted as an intensification of some traits of modern rule. This is an important point. Making sense of neoliberalism and its relationship with science and nature means paying attention to gradual and subtle alterations of established practices and ideas, rather than searching for major differences or revolutionary changes.

The Foucauldian perspective is especially committed to connecting neoliberalization with an intensification of established views, which leads to a new understanding of human nature and social existence; a novel conception of how society is to be governed and social relations defined (Read 2009, Tazzioli 2011). While liberalism assumed a natural tendency of humans to exchange and saw the market as a self-regulating institution, neoliberalism assumes a natural tendency of humans to compete within task environments that have to be purposefully constructed, steered and policed (Burchell 1996, Tickell and Peck 2003). Competition, in other words, is the essence of human rationality, but at the same time this feature has to be actively supported. Individual rivalry has to be stimulated and sharpened, expanding as much as possible the diversity of positions, the 'equal inequality' of people (Foucault 2008, see also Lazzarato 2009). The basic task of neoliberal policies, therefore, is to develop, disseminate and institutionalize economic rationality in any social field (Brown 2005: 40–41). Everyone is (to become) an entrepreneur of themselves, responsible for their own choices. Illness, unemployment or poverty become one's own responsibility and failure (Lemke 2004, MacLeavy 2008). As a form of subjectivity, 'human capital' replaces the free labourer of the liberal tradition. The distinctions between production and reproduction, public and private, professional and domestic spheres are eroded, and the values and attachments in which market calculations and labour were traditionally embedded are increasingly subsumed to an entrepreneurial logic (Feher 2009, see also Bárd this volume).

Entrepreneurship is simultaneously a matter of rationality and a moral duty. Truth and ethics, moral responsibility and economic rationality, overlap in the free action of individuals based on cost-benefit assessments. The scope of such assessments expands together with technoscience advancement, as with the 'new ethics of biological citizenship and genetic responsibility' (Rose 2007: 39) that emerges as a consequence of the advancement of biotechnologies. As a result, government operates more on interests, desires, and aspirations than through rights and obligations. 'The regulation of conduct becomes a matter of each individual's desire to govern their own conduct freely in the service of the maximization of a version of their happiness and fulfilment that they take to be their own' (Rose 1996: 57). People are oriented in an increasingly indirect and pervasive way towards their self-fulfilment or enhancement.

All this, as objective thought forms expressed in discourses and institutionalized in norms and rules, profoundly affects the way in which human agency is understood. The neoliberal *homo economicus* is not calculative in the same sense as the liberal one. The liberal view of freedom, rationality and responsibility entails a future neither totally fixed nor totally random. Risk means a future event related to behavioural choices, the probability of which is amenable to calculation. At the same time, non-calculable uncertainty prevents humans from being prisoners of an inevitable path. Real profits, economists like Keynes and Knight tell us, stem from 'unpredictable risks', related for example to innovation, which are the object of a few strategic decisions. In this sense, the complexity of climate and ecosystems cannot but frustrate the liberal agent's purposeful action.

For neoliberalism, however, uncertainty 'makes free' in a different, more 'intense' way. Decision-making under uncertainty becomes an empowering everyday situation.

> An extensive and immensely influential managerial literature appearing since the early 1980s [...] celebrates uncertainty as the technique of entrepreneurial creativity, [...] the fluid art of the possible. It involves techniques of flexibility and adaptability, requires a certain kind of 'vision' that may be thought of as intuition but is nevertheless capable of being explicated at great length in terms such as 'anticipatory government' and 'government with foresight'. (O'Malley 2004: 3–5)

Entrepreneurial agency is located within the artificially arranged, ever-changing task environment produced by global trade, innovation-based competition and the financial turbulence created by floating exchange rates. This impinges on how contingency is accounted for. Indeterminacy does not mean constraining *non-determinability*, but enabling *non-determination* (Pellizzoni 2010). Whilst in the former case the causal chains are regarded as open in the sense that the events can take unpredictable turns because of unknown intervening variables, emergent systems properties and so on, in the latter case the causal chains are regarded as open in the sense that the agent does not find them predetermined, but can handle and orient them in the desired direction. Contingency means lack of limits, room for manoeuvre. The more unstable the world, the more manageable it is. Rather than paralysing, the eventuality of the future, or the subjectivity of expectations, enables the construction of purposefully designed task environments where new opportunities take shape. Neoliberal agents estimate 'the future in much the same way that people do engaging in extreme sports: that is by accumulating information, relying on experience, using practiced judgment and rules of thumb' (O'Malley 2008: 73). Their orientation is speculative rather than predictive: proper calculations of risk are seen as the exception, while reasoned bets on unpredictable futures are regarded as the rule. Whence derives the growing role of hype, anticipations, expectations, imaginaries, scenarios in politics, economy and technoscience (Brown and Michael 2003, Felt and Wynne 2007, Cooper 2010).

Neorationality

Neoliberalism, it has been pointed out, 'is a profoundly active way of rationalizing governing and self-governing in order to "optimize"' (Ong 2006: 3). The above discussion, however, indicates that this 'optimization' does not operate according to a traditional understanding of rationality. It is therefore important to clarify where precisely the difference lies.

In modern times the relationship between human agency and the biophysical world has been understood in three basic ways. The first two are realism and epistemic constructivism. We have either a direct access to, and control of, an ontologically given reality, or access and control are mediated by knowledge, perception, value commitments and social organization. Both positions contrast with the ontological variants of constructivism that emerged within post-structuralism. For many scholars in science and technology studies, 'knowing, the words of knowing, and texts do not describe a pre-existing world [but] are part of a practice of handling, intervening in, the world and thereby of enacting one of its versions – up to bringing it into being' (Mol and Law 2006: 19). Compared to realist or epistemic constructivist accounts, this view entails a much 'weaker' understanding of human agency by which objects come into existence together with the discursive formations that make it possible to talk about them. Better, we observe a double weakness. Both the world and we who act upon it are ontologically plastic. Humans affect the world and at the same time the world affects them (Szerszynski, Heim and Waterton 2003, Wynne 2005).

Neoliberal rationality cannot be likened to any of the above positions. On the one hand, as regards the ontology of the biophysical world, it is resolutely constructivist. On the other, as regards agency, it argues totally to the reverse of post-structuralist scholarship. If everything is intimately related, states the latter, everything has to be treated carefully, respectfully, with prudence and restraint. If everything is crafted, the neoliberal reply goes, everything can be remoulded, commodified and appropriated. World-making does not mean a joint constitution of subject and object, but a purposeful manufacturing of the latter. Nature is pliant; will is unconstrained. Unconstrained, of course, in its potentiality. As act, application to an object of desire, it cannot but be sensitive to a context that, as said, is increasingly characterized by the indirect guidance, discreet suggestions, imperceptible orientations provided by a variety of cultural pressures and expertises (see also Bárd this volume).[6]

The core of neoliberal rationality, therefore, lies in the interplay between world plasticity and strong (albeit 'conducted') agency – something that, to our knowledge, has gone largely unnoticed in the governmentality literature. We call this *neo*rationality, since we are not faced by a farewell to modern reason, like the post-modern farewell to grand narratives and objectivist theories of knowledge, but rather by a double-edged intensification of the latter (Pellizzoni 2011a). One pillar of modernity is abandoned: the core distinction between inner and outer world disappears in favour of what, to all intents and purposes, is an anti-essentialist ontology. At the same time, another pillar of modernity, traditionally linked to the idea of objective knowledge, is reaffirmed and expanded in its scope: human agency as having capacity of control; an agency that finds no 'external' limits since it includes the manufacturing of its own task environments. Limits, therefore, pertain only to the agents' horizons of meaning, which mould their understanding of what can and should be done in the context of the neoliberal

6 A recent variant of this orientation is the concept of 'nudge' proposed by Thaler and Sunstein (2008), who make a case for a 'libertarian paternalism' by which, through appropriate techniques, public and private organizations should gently and imperceptibly forge people's habits, helping them make the 'right choice' (for their health, wealth and happiness).

intensification of the modern values of progress, overcoming, achievement: more wealth, health, beauty, intelligence, physical smartness, 'naturalness', and so on.

This double-edged rationalization represents, in our view, the ideational engine of neoliberal hegemony and a key to making sense of its relationship with technoscience. It is one thing to optimize performances within an ontologically stable world (or relinquish any such aim within an ontologically unstable one); it is another to optimize them with respect to an ontologically plastic reality. Liberalism was concerned with the material limits to growth. Neoliberalism is concerned with the growth of limits (Lemke 2003). The pressing search for new fields of expansion and acquisition finds an indispensable ally in technoscience: nature is no longer an ultimate barrier but a moveable threshold. The consequent expanded accumulation is then managed through the production and allocation of scarcity by means of property rights and markets (Harvey 2003).

The neorational combination of strong agency and world pliancy makes the old distinction between knowing and intervening obsolete, yet in a different way to post-structuralist thinking. There is a crucial difference between the pre- or non-modern world described for example by Latour (1993) and the neoliberal world: the intimate connections of words and things, facts and concerns, are now handled by an agent that seems anything but weak, being committed to producing contingencies in a purposeful, manageable way. As Foucault (2008) remarked, neoliberal entrepreneurs are likely to be only temporarily limited by their biophysical constitutions; technoscience is bound to overcome such limits.[7] This gives a clue to the actual character of the neoliberal agent. Contrary to widespread readings, the latter has less to do with *homo faber* and his Promethean hubris than with the *homo creator* described by Günther Anders (2003): a human who is able to generate from nature products that do not belong to culture but to nature itself, overcoming any essentialism by means of the unconstrained expansion of a creative will extroverted beyond any discernment about its own goals.

Scientism and ontological commodification: The case of gene technologies

Technoscience is thus profoundly traversed by the neorational logic of neoliberalism. The process is two-sided: on the one hand, neoliberal government includes growing amounts of 'science'; on the other, science is increasingly shaped according to neoliberal rationality. We may call the former process *scientism* and the latter *ontological commodification*. Gene technologies represent an elective terrain for analysis of both.

---

7 See for example the influential policy narrative of 'converging technologies': that is, the alleged ongoing combination of nano-bio-info-cogno technoscience towards an unprecedented 'enhancement' of industrial productivity and human biological and mental capacities (Roco and Bainbridge 2002). On this point see also Arnaldi (this volume) and Bárd (this volume).

Scientism can be defined as 'the belief that politics is best dictated by scientific reasoning' (Kinchy, Kleinman and Autry 2008: 156); that 'sound science' transcends values and interests providing objective answers to policy problems on which all parties can agree (Moore et al. 2011). Neoliberal scientism intensifies the traditional focus of the modern rationality of government on knowledge and truth claims. On the one hand, the need for stronger international standards stemming from the reduction of trade barriers and the increased internationalization of markets gives technical bodies a growing regulatory importance: consider IPCC, ISO or the Codex Alimentarius Commission. On the other, governmental action is increasingly focused on 'risk', that is, on the safety, quality and efficacy of innovation beside considerations of need, desirability and distributional effects. The regulation of the recombinant bovine growth hormone (rbGH) or the Cartagena Biosafety Protocol on the transport and use of genetically modified organisms provide examples of how socio-economic aspects like the impact of innovation on traditional farming and small producers are systematically downplayed (Kinchy, Kleinman and Autry 2008).[8] Indeed, a long-standing orientation of the European Union is that adopting a more encompassing perspective 'could result in a diversion of investment and could act as a disincentive for innovation and technological development by industry' (CEC 1991: 8). Only ethical issues (e.g. for animal welfare) receive some consideration.

Neoliberal hegemony surfaces here in a refined form. The integrative function performed by science (and ethics: see below) legitimizes a biased, depoliticized portrayal of the issues at stake (general benefit of innovation) which excludes any non-scientific (or non-ethical) arguments as incongruent or 'ideological'. For concerned groups, therefore, a pragmatic acceptance of this justificatory plane may represent the only alternative to marginalization.

We have here also an example of how neorationality permeates concrete regulatory choices. On the one hand an active protection of small farmers would mean intervening directly in market dynamics, whereas the authority of science helps to establish a 'neutral' framework for competition. On the other, the appeal to sound science has little to do with an old-fashioned objectivism and much more with an anti-objectivist attitude, since it can be instrumental to opposite purposes. For rbGH or GMOs, sound science has been claimed to provide sufficient evidence of no problems, triggering green lights for the new technologies. Yet in the case of climate change, for example, sound science has been claimed to provide no sufficient evidence of problems, triggering red lights for restrictive measures (Freudenburg, Gramling and Davidson 2008).

Science, however, is not only *used*; it is also *shaped* according to a neoliberal logic – as a reflection on biotechnology patents highlights. As a resource, knowledge represents an elective terrain for neoliberal rationality. It is non-rival (its use is non-competitive) and can be reproduced at a much lower cost than the

8 A similar situation can be found in the regulation of biofuels: see Levidow, Papaioannou and Birch (this volume).

cost of its production. Rivalry, as a consequence, has to be constructed. Patent theory maintains that by assigning property rights – that is, by transforming knowledge into a scarce, marketable commodity – patents promote the creation and diffusion of innovation. Biotechnology patents thus expand the proprietary character of knowledge to the detriment of the intellectual commons (Jessop 2010). Moreover, patents require novelty, inventiveness or non-obviousness, and applicability. This in principle excludes pure information (ideas, theories etc.) and things or processes that can be found in nature. However, the purification of a natural substance has long been recognized as patentable (Carolan 2010a). In short, biotechnology patenting can be regarded as an intensification of existing patterns in the regulation of innovation. The result is the qualitative change we call ontological commodification.

The story of biotech patents is well-known (see also Reynolds and Szerszynski this volume). In its famous 1980 ruling on *Diamond v. Chakrabarty*, the US Supreme Court stated that a genetically modified bacterium is human-made, that it is a new composition of matter, and that whether an invention is alive is not a legitimate legal question. In short, 'anything under the sun made by man', including living matter, can be patented. Precisely in the same year, the Patent and Trademark Amendment (or Bayh-Dole Act) was passed in order to promote the patenting of publicly funded research and the private exploitation of patents by their holders (either by issuing licenses to private companies or by entering into joint ventures, or else by creating their own start-up companies). The novel and by now famed figure of the scientist-entrepreneur stems from this legislative innovation. The world-wide spread of this regulatory approach was subsequently ensured by the 1994 TRIPs (Trade Related Aspects of Intellectual Property Rights) Agreement, strongly advocated by North-American pharmaceutical, software and entertainment industries. TRIPs extended the scope of patentability to all commercially exploitable products and processes, including genetically modified plants and animals. Ratifying TRIPS is mandatory for joining the WTO, so that every adhering state is compelled to adapt its own legislation accordingly.

The Supreme Court or the political and corporate supporters of the Bayh-Dole Act and the TRIPs Agreement may not have been directly inspired by neoliberal theories.[9] Yet, as Foucault reminds us, power and domination are matters of actual effects, not of agents' goals. Effects, in this case, consist in the enforcement of a particular understanding of human relationship with nature. A product patent for a genetic sequence, for example, entails regarding it as a 'composition of matter', novel in that it is not available in nature in its isolated and purified form, and whose 'utility' or 'industrial applicability' lies in the disclosure of its function. Such disclosure basically corresponds to understanding the biochemistry of the protein that a gene produces and how this engenders a specific trait of the organism. Genes are therefore regarded as carriers of information suitable for

9 Concerns about declining profits, national competitiveness and the handling of technology transfer arguably played a major role (Cooper 2008, Moore et al. 2011).

translation into different media (Kay 1999). Though, as said, information as such, like ideas or scientific theories, is excluded from patenting, the demonstration of some technical effect or functionality allows for property rights claims. This entails disregarding the often complex connection between genes and traits: one gene may be involved in the production of many proteins and there are typically several molecular interactions, cascades and feedback loops responsible for the final phenotype (Calvert 2007).

We can see here again the traits of hegemony: the integrative function performed by patent regulation simultaneously legitimizes the bracketing of distributional issues, to be addressed through the market, and a biased account of scientific evidence. We can also see the traits of neorationality. Patents carve out their own task environments, transforming indeterminacy from a drawback into a valuable resource. On the one hand any difference between living and non-living entities is erased. On the other a living entity is considered an artefact if basic functional parameters can be controlled, thus reproduced, and a correspondence is implicitly established between matter and information, so that rights in property over information can be subsumed into rights in property over the organisms incorporating such information, and vice versa. This ontological ambiguity or oscillation translates into actual court rulings.[10]

Hence nature is what provisionally belongs to the ill-clarified, non-domesticated world lying beyond the boundaries of commodification set by commodity producers. It is the 'environment' of the capitalist system (Lemke 2003). Better, nature and manufacture become distinctions internal to the manufacturing process (Pellizzoni 2010). Biotechnology patents are 'fluid objects' (Carolan 2010a; see also Carolan 2010b).This fluidity also results from the 'substantial equivalence' argument central to commercial applications by which, for any practical purpose, patented artefacts are indistinguishable from nature; they consequently do not require any specific regulation. Artefacts are thus simultaneously identical to and different (more usable, more valuable) from natural entities.

With gene technologies, nature becomes artificial and culture becomes natural (Rabinow 1996). Yet this ontological fluidity is shaped according to a market framework. It is therefore important to grasp where the real novelty lies. Capitalism has traditionally treated elements of nature, such as water or trees, as 'fictitious commodities': that is, marketizable resources disembedded from their socio-cultural meaning and biophysical function (Polanyi 1944). Here, however, we do not have an epistemic abstraction. There is more than an 'as if' at stake with patented genes: there is the actual crafting of entities that did not exist beforehand and that are characterized by a structural ambivalence: equivalence-difference, materiality-virtuality, substance-information. There is nothing fictitious about these commodities: they *are* commodities, their 'reality' is nothing other than this. In short, biotechnology patenting expresses neoliberal

10 See e.g. the 2004 *Monsanto Canada Inc. v. Percy Schmeiser* case on the use of the Roundup Ready™ canola plants, discussed in Carolan (2010a).

neorationality at its best: construction of market competition, intensified agency, indeterminacy as a resource, ontological commodification as a result. Nature, it has been noted, is progressively subsumed to capital in a 'real' rather than 'formal' way. Instead of being used according to their intrinsic features, natural resources are increasingly altered in order to enhance their productivity (Boyd, Prudham and Schurman 2001). Yet this account seems more suited to non-genetic industrial biotechnologies. With gene patents, more than an alteration, we have the creation of resources that belong to nature but at the same time are totally internal to the economic process of accumulation. We can talk of real subsumption, then, only by stretching its meaning to encompass not only an enhancement of the productivity of natural resources but a profound redefinition of their ontology, according to market logic.[11]

Remoulding nature, remoulding humans

By no means are these special features of the biotechnology field, since the same rationality can be found elsewhere, for example in carbon trading (Pellizzoni 2011a; see also Blok, this volume). What this field highlights, however, is that the 'neoliberalization of nature' includes a remoulding of humans in their biological and social identity. We focus here on two issues: 'clinical labour' and 'benefit-sharing'.

Genomics entails the application of information technology to study the interactions among genetic variations, environmental and lifestyle factors, and the aetiology of diseases. For this purpose, researchers need access to biobanks, that is, broad, systematic collections of human biological materials, but also to the participants' bodies in their living experience. Clinical labour, therefore, is 'the regularized, embodied work that members of the national population are expected to perform in their role as biobank participants – in the creation of biovalue through biobanks' (Mitchell and Waldby 2010: 334).

Biobanks raise issues of privacy, confidentiality, consent, trust, citizen rights, civil participation and so on. Yet Mitchell and Waldby stress further, and possibly deeper, implications. First, a tendency to connect the public good with the development of commodities in the form of drugs and diagnostic tools. This not only because biobanks provide data to the pharmaceutical industry, but also because genomic research in itself is more attuned to developing diagnostic technologies and pharmacokinetic studies than basic aetiological studies. Second, biobanks are focused less on the genetic causes of diseases than on risk factors for disease. Individuals with diseases are linked to their genetic profile and environmental exposures, with the goal of distinguishing high-risk from low-risk genotypes.

11 This process can be seen in close connection with Reynolds and Szerszynski's (this volume) account of the novel form of subsumption of science to capital under neoliberal rule.

Third, as noted, biobanks entail access to both the in vitro and in vivo biology of people. This includes compliance with medical regimes of dosing, testing, self-monitoring, and more in general, checking or adjusting one's lifestyle. The entire lives of participants are involved in a productive relation with biobanks.

If we put these features together, the traits of neorationality are clearly detectable. On the one hand, as the notion of biovalue suggests, we are confronted with an overarching economic logic of (self-)investment that connects people, the public good and commercial interests together. On the other, genetic analysis carves out task environments that transform uncertainty from a limit into a resource. Risk factors are relational categories; their logic is associative. The search for risk factors is probabilistic in a strict sense only for a limited range of diseases. In most cases, estimations are made on the basis of associations among heterogeneous elements which diagnostic devices translate into actual realities.[12] We are once again confronted with ontologically indeterminate entities. Finally, people's own ontology is redefined, becoming ambivalent and opaque: participants in biobanks are simultaneously tissues and genetic information, individuals and risk categories, autonomous and heteronomous subjects. The creation of biovalue needs the pliancy of human bodies and minds.

Participation in biobanks is usually depicted as a way of contributing to the common good, an act of altruism, a 'gift to strangers' similar to blood donation. To many, however, altruism and donation seem at odds with clinical labour as value production. Participation as gift-giving is increasingly challenged by the alternative notion of 'benefit-sharing'. The idea is that 'participants in research deserve some form of returns, precisely because their participation is leading to lucrative products' (Hayden 2007: 731). This perspective applies, for example, to bioprospecting activities: the collection of forms of life and related 'traditional knowledge' for their scientific and commercial potentials. According to the 1992 UN Convention on Biological Diversity, such forms of life and culture cannot be considered part of the global commons. Therefore, source communities and nations are entitled, in an ethical rather than legal sense, to have some share in the value that they generate. The principle is clear, even though specifying the form (monetary or otherwise) and magnitude of the returns, and their beneficiaries (local residents, ethnic groups, national researchers) can be problematic.

In clinical research, benefit-sharing connects the growing evidence that biological samples and their derivatives are lucrative forms of property for companies with the emergence of collective research subjects or other forms of 'biosociality' or 'biomedical citizenship' (Rabinow 1996; Rose and Novas 2005): patient and 'high-risk' groups, relatives of diseased persons, volunteers for trials,

12 Another field in which this logic receives growing application is surveillance. Risk profiling techniques and biometric algorithms are increasingly applied to generate probabilistic rules of association among disparate elements (persons, places, behaviours, events); associations that are subsequently interpreted as indicators of 'threat' (Amoore 2009). On this point see also Bárd (this volume).

ethnic groups, entire nations (for national biobanks). An additional problem is the need to prevent 'undue inducement' to participate in research. The thin line between unethical inducement and ethical benefit-sharing increases the importance of 'what' is given back and 'to whom', and the perceived need to avoid any legal drift (attribution of property rights to one's own body and thus to the results of research).

Benefit-sharing represents another point of emergence of neorationality. Not only is the overarching economic logic of the whole issue evident, but it follows a typical pattern of regulation. It should be borne in mind that the neoliberal goal is not to marketize everything (as would happen in our case by attributing property rights to body parts and connected biovalues), but rather to create the conditions for market exchanges, which entails that non-market relationships be aligned with a market logic. Whatever its concrete specification, benefit-sharing challenges the traditional gift-giving approach to the use of body parts, signalling the ceaseless expansion or intensification of neoliberal rationality. However the recipients and the character of the benefits are defined, redistribution is privatized: 'there are no strangers in this vision, only fellow (and competing) stakeholders' (Hayden 2007: 749). The ethical framing of the issue (equity vs. property as a guiding rule) is therefore misleading. Apparently following an established legitimation pattern of bioscience research and working as an antidote to the full commodification of bodies, knowledge and biodiversity, ethics effectively acts as a legitimating mediator between science and the market. Conspicuous by its absence is the political aspect of the issue: the power relations inbuilt in the deal and their distributional effects. The problem of the recipient communities' identification is exemplary of the displacement of the political. These collectives are treated as already-existing ethical subjects, yet they are actually constituted in the encounter of clinical 'labourers' with scientific-corporate 'employers'; a constitution that represents a classic political gesture of inclusion and exclusion. Such constructivism, moreover, exhibits marked neorational features. The ontology of these collectives is hybrid in that it oscillates between nature and culture, genes and habits, biological condition (disease) and social relation (care), the sharing of a physical place and the sharing of a role in the production chain. This is an indefiniteness that, once again, does not hamper but instead enables purposeful action.

Neorationality and depoliticization

If depoliticization is intrinsic to all hegemonic projects, especially in the framework of modern governmentality (replacement of people with populations; blurring of the public and private, political and economic, etc.), what is relevant here is the neoliberal intensification of this process in the technoscience field. We have already seen the leading role played by scientism; yet the place of 'ethics' and 'dialogue' warrants further discussion.

The spread of ethics councils can be regarded as the institutional counterpart of the spread of the enterprising, responsible 'ethical citizen'. Presented as 'a "neutral" normative tool, endowed with the potential to speak for rationality' (Tallacchini 2009: 281), these councils are composed of appointed 'experts' allegedly able to represent relevant viewpoints and concerns, or, in the case of public citizen dialogues, to interpret inputs from, and give proper guidance to, the reflections of 'lay' people. Ethics councils, moreover, explicitly refrain from addressing political issues.[13] Discussions focus on questions of 'how', rather than 'whether or not' or 'for the benefit of whom'. Ongoing innovation and its contribution to the welfare of society as a whole are taken for granted (Braun et al. 2010). Ethics is thus framed as setting appropriate limits to technoscience's advancement; yet these limits are continuously expanded by that advancement itself. The trading and exchange of values is normalized and legitimated: 'Through the enunciation and application of a set of principles, standardized rules are established that enable the translation of different moral positions to a common metric capable of facilitating, usually on a cost-benefit basis, choices and decisions' (Salter and Salter 2007: 560). If possible, counter-narratives are included as legitimate differences, as distinctions internal to the agreed framework. If non-negotiable positions emerge, they are marginalized or expelled from the regulatory process. When lay people are engaged in ethical debates, the task is to promote 'proper talk' (Braun et al. 2010): that is, an 'appropriate' style of discourse based on reflexivity, moderation, and openness to compromise. Again, non-negotiable standpoints are marginalized, often being stigmatized as ignorance or prejudice.

For some, ethics has become the 'decisive semantic form' (Bogner and Menz, 2010: 890) of technoscience governance. With ethics councils, at least, we observe a 'politics of ethics' (Felt and Wynne 2007: 47) where neorationality is fully operationalized in its radical constructivism and strong agency. On the one hand, the object of discussion – for example, stem cells – oscillates between a scientific and an ethical ontology, so that scientific arguments and approaches are used to legitimize ethical ones, and vice versa (Salter and Salter 2007). On the other hand, the indeterminacy of the relevant values and of their regulatory translation is turned into a resource by creating 'workable' task environments where antagonistic views are recodified as variations internal to a shared vision, or ignored.

Another keyword of neoliberal governmentality is 'dialogue'. True, deliberative democracy has been advocated by Habermas and a number of scholars on both sides of the Atlantic as precisely an antithesis to rational choice politics and 'public management' policies of privatization and individualization of services

13 Their rise, in fact, corresponds to the decline of different, more politically-minded, approaches to expert contribution to regulatory processes The most noteworthy case is that of the US Office of Technology Assessment. Operating the early 1970s onwards and closed in the mid-1990s, OTA had the mandate to address physical, biological, economic, social and political (but *not* ethical; at least, not as a specific, isolated issue) aspects of technoscience.

and responsibilities. Yet there is no denying that managerial-technocratic policy styles have made large recourse to public deliberation at different scales, from EU regulation to urban planning. More in general, deliberative arenas began to spread in the 1990s, precisely at the moment when neoliberalization entered its 'roll-out' phase (Tickell and Peck 2003), with a series of measures aimed at countering the dramatic socio-economic consequences of deregulation, marketization and the dismantling of the Keynesian state. What is most indicative from the viewpoint of hegemony, however, is the focus of *both* deliberative democratic and public management discourses on citizen 'empowerment' and policy-making as problem-solving. Let us briefly consider some evidence emerging especially from research in the environment and technoscience field.[14]

First, deliberative processes are framed according to a dominant 'stakeholder' language. 'The general public or the community [is] portrayed as one of a number of stakeholders in some sense entitled to be party to the dialogue' (Goven 2006: 104). Second, deliberative arenas are designed and enacted from above. They produce a 'governed' empowerment of citizens. The assumption is that fruitful, constructive discussion needs a definite task, appropriate rules and a suitable number of the 'right' participants. Specifying these aspects is presented as a technical matter of efficiency, rather than a political act of inclusion and exclusion. Third, participants are gathered as 'proxies' for the general public and tasked with reflecting on the best (or a viable) solution to a 'problem'. This entails a consensual process through which the principled commitments of lay participants are matched with the factual evidence (or reliable anticipations) offered by 'science' as expert-sourced information or direct advice (Wynne 2001). Fourth, political judgements aimed at defining the *general* interest (which inevitably means making some interests prevail over others, albeit in a publicly justified manner) are replaced by a sort of judicial assessment aimed at impartially defining the interest of *all* (Urbinati 2010). Fifth, it is first and foremost in the 'internal forum' that this assessment is assumed to occur (Goodin 2000): a mental process of clarification about facts and values and their harmonization for the common good, which balanced information and appropriate rules of talk should protect from undue influences.

Evaluations of specific experiences of course vary.[15] Yet, overall, deliberative arenas can be critically regarded as ingenious ways of conducting conducts.

---

14 That deliberative processes have found in this field an elective terrain is a further testimony to the centrality of science, technology and nature for neoliberal policies and the connected struggles.

15 For example, commenting one of the most conspicuous experiences, the 'GM Nation?' public dialogue organized in the UK in 2003, Irwin remarks that 'in giving the appearance of democracy, such talk actually diverts from a more adequate onslaught on deeper institutional and epistemic commitments [...]. Little has changed: we are simply in the old nexus of technocratic aspirations with the public construed as an obstacle to progress' (Irwin 2006: 316–17). For others, however, this very experience shows that deliberative arenas can also offer an opportunity for challenging boundaries, performing

These processes, moreover, are profoundly neorational. Deliberative arenas are 'machineries for making publics' (Felt and Fochler 2010), the members of which must project an image of the general public onto strongly bounded issues and with the crucial assessment occurring in their own internal forum.[16] Deliberative arenas are purposefully designed task environments that transform into resources what would be puzzling indeterminacies: about the subject, object, addressees and outcome of deliberation. They craft entire worlds furnished with people, problems, solutions, facts, values, and handled by individual agents that, while committed to searching for the common good, are led to reproduce the traits and moves of the neoliberal entrepreneurial agent.

Hegemony, neorationality and resistance

The case for resisting neoliberalization raises three analytical questions: why resist, how to do so, and who are the agents of resistance? Each of them is affected by the peculiarly hegemonic character of neoliberal rationality.

There is no shortage of criticisms against neoliberal politics and policies. They usually point to inability of the latter to match the 'lofty ideals' of neoliberalism (Heynen et al. 2007, Robinson 2004) by delivering the promised goods (more environmental efficiency and economic stability; more wealth, health, well-being, freedom for all), and to their disrespect for core values of political modernity: equality, substantive citizenship, solidarity, defence of common goods against appropriation, respect for nature, and so on (Robinson 2004, Brown 2005, Harvey 2005 and 2006). The persistence and resilience of neoliberalism is accounted for in terms of both institutional flexibility (Brenner, Peck and Theodore 2010) and the 'ecological dominance' of a world-wide, market-mediated logic of capital accumulation over other ensembles of social action (Jessop 2010). Global capitalism, in other words, is hegemonic not only because its legitimizing basis (neoliberalism) integrates a variety of positions while obscuring its biased account of social affairs, but also because it has the ability to provide material rewards and to impose sanctions (Robinson 2004: xv). Adaptation to dominant ideas is the result not only of their intrinsic force but also of actual power relationships, crystallized in political mechanisms and institutions which enhance ideas and firmly entrench them (Abercrombie et al. 1980, Heynen et al. 2007). The growing import of corporate power (Crouch 2011) is hardly irrelevant in this respect. In the biotechnology field, as we have seen, regulatory processes, bioprospecting and clinical research propound particular views of science, the public interest and

different models of the public and questioning dominant expert assumptions (Levidow 2007). An assessment of deliberative democracy in environmental politics is provided, among the others, by Backstrand et al. (2010).

16 Constructed or imagined publics play a major role also in the decision-making of actors in technical-industrial and policy networks (see e.g. Walker et al. 2010).

people as ethical subjects, which depend on and simultaneously justify biased narratives and asymmetrical roles among citizens, scientists, corporations and political authorities.

Yet neoliberal hegemony possibly finds a crucial factor in the specific character of neoliberal rationality. A truth regime can be called into question by 'unformattable' accumulating evidence; things that do not fit with its truth-false, reasonable-unreasonable classification criteria. Humans test ideas by acting: gathering contrasting evidence may lead to the questioning of objective thought forms. However, as we have seen, neoliberalism is a political project that seeks to create a social reality that it maintains already exists (Lemke 2003). It develops institutional practices and rewards in order to expand competitive entrepreneurship. Simultaneously, however, it claims to present 'not an ideal, but a reality; human nature' (Read 2009: 26). If competitive rationality constitutes the fundamental character of human nature, there is no empirical evidence able to confute it. Any failure of the market, any evidence opposed to the promised increase in freedom and efficiency simply marks the distance between a trans-historical reality and contingent flaws, constraints, oppositions and irrationalities. True, any hegemonic order seeks to reject contradictory evidence as irrational behaviour or deficient practice, trying to naturalize particular task environments suitable for optimizing human behaviour according to a given model. Yet neoliberalism 'intensifies' these features. Description and prescription, actual reality and teleology, are completely conflated. The constructed character of entrepreneurial task environments is openly celebrated, while their natural character is simultaneously reasserted. It is this ontological oscillation, on which we have insisted throughout the chapter, that gives neoliberalism its special resistance to confutation and contestation.

More than on a discursive level, then, resistance should build on praxis – concrete action. This is suggested by dialectical materialism, as well as by Foucault, with his stress on counter-conducts as a major site of resistance. Here, however, we meet other problems. Foucault warns that there is no power without resistance. At the same time there is no 'outside' to power that could check power: resistance is an effect of power. Its emancipatory potential therefore lies in its capacity to 'overhang' power. It lies in the unpredictable dynamics of power relations, the incompleteness of domination. Similarly, hegemony theory maintains that the adequacy of objective thought forms to actual social experiences is always limited, so that every hegemonic project is incomplete (Hall 1983). Then, as power 'becomes increasingly more capillary, more invested in everyday matters and everyday lives, so too an immense new field of possibility for resistance is opened' (Nealon 2008: 107–8). This is the great opportunity offered by current governmental rationality. For the late Foucault the elective terrain of 'practices of freedom' is actually ethics: the 'care for the self' (Foucault 1997a). Yet we have seen that the expansion of ethics in the governance of science is problematic because of its depoliticizing implications and its alignment with the neoliberal self-enterprising model of the human agent. By encouraging individualized reflexivity, ethicization can discourage 'associative relations among individuals and attempt[s] to contain the "counter-power" potentially generated by associational activity' (Myers 2008: 128).[17] The ethical terrain makes it easier to adjust to criticisms and concerns without threatening the established order.

This ambivalence invests science also as a field of open conflict. Objective thought forms are increasingly based on science and expert knowledge; as a consequence, the importance of science in counter-hegemonic struggles has grown. Scientific methods and languages are increasingly applied in protests. Activists rely on sympathetic scientists or undergo training to acquire a sufficient degree of expertise (Moore et al. 2011). Neglected issues are brought to the fore and counter-expertises are built up (Frickel et al. 2010, Pellizzoni 2011b). Technical discussions may have political effects to the extent that they open spaces for contestation of the underlying political choices (Barry 2002). However, grounding contestations on a scientific terrain (or giving this type of argument a pivotal role) means also aligning with the neoliberal treatment of technoscience governance as a mere technical issue. The net result in terms of the capacity of resistance to 'overhang' hegemonic forces is an empirical question, to which no clear answer is available.[18]

From a Gramscian viewpoint, counter-hegemonic constellations emerge continuously and everywhere. People's material positions and everyday experiences define what elements of hegemonic discourses and objective thought forms are accepted or discarded; yet the precondition for effective resistance is that fragmented views and experiences be connected together. How this may happen in a globalized world where political steering is increasingly mediated by networks of unaccountable technoscientific and economic agents represents a major challenge.

The Foucauldian perspective is of little help here. It claims that subjects are free by definition and resistance is intrinsic to power. Resistance, therefore, cannot be produced from scratch. Rather, it is necessary to find 'ways to mobilize, focus, or intensify, practices of resistance, insofar as they are already all over the place' (Nealon 2008: 105). Yet when it comes to being more precise about this

---

17 Also from a Marxist viewpoint, what is effective is *collective* praxis: the experiences and questionings of single individuals are not sufficient to change objective thought forms (Hall 1983).

18 See the already-mentioned discussion by Kinchy, Kleinman and Autry (2008). On similar lines, Ottinger (2010) finds that focusing on quality standards may support but also limit the scope of citizen mobilization against pollution. Analysing a controversy concerning mobile phone masts Drake (2010) reaches an analogous conclusion: health and the related scientific uncertainties represent a theme that is both publicly resonant and aligned with the neoliberal discourse of individual risk and responsibility within a framework of inescapable and benevolent market-mediated technological progress. People are simultaneously expected to increase their responsibility-taking and accept the constraints of technical and commercial imperatives. This contradiction emerges only when the whole framework and not only the contours of 'sound science' are brought into question.

'tuning', the Foucauldian literature is vague.[19] On the other hand, as hinted, for hegemony theory consent derives from not only normative but also pragmatic acceptance. This is important. If people align with governmental requirements not only because they believe them right and effective but also – as the example of biobanks and many others indicate – because they have no alternative, then neoliberal 'government at a distance' may prove to be less pervasive and encompassing than many Foucauldians argue. Moreover, the problem of the shape and size of counter-hegemonic forces can hardly be dismissed, even in Foucault's own terms. For him, governmentality does not replace but integrates sovereignty. Therefore, resistance must simultaneously address governmental and sovereign powers, indirect regulatory strategies and direct (sometimes violent) commands.

This twofold character of resistance is reproduced in the present division of the neo-Marxist camp into two factions. For Hardt and Negri (2004) and Virno (2004, 2009), the subject of resistance today is not the class of the old capitalist mode of production, or the people of the old nation state, but the 'multitude': 'a middle region between "individual and collective"; [...] *a plurality which persists as such in the public scene*' (Virno 2004: 21, emphasis in the original). As a network of singularities, the multitude (see also Gandini, this book) is the political translation of the basic productive force of the knowledge-based economy: a mass intellectuality which constantly elaborates and applies thoughts and discourses, cognitive competences, imagination, as the general capacity of reflection and communication that presupposes a common participation in the 'life of the mind'.

This new type of subject 'presents itself on the public stage as an *ethical* movement' (Virno 2009: 103, emphasis in the original). The new forms of sociality emerging in the biotechnology field might then be taken as salient manifestations of the multitude. Yet this is precisely what other scholars find problematic. Traditional politics and the state as a sovereign regulatory and military entity, it is stressed, have lost none of their relevance, requiring commensurate counter-forces which can hardly be represented by the unconnected, ad-hoc, ethically-minded mobilizations of the multitude. Biosocialities and single-issue deliberative arenas are regarded with suspicion, together with appeals to 'rethinking politics', because of their alignment with neoliberal arguments (Mouffe 2005, Jessop 2010). Even the growing involvement of NGOs as policy advisors or delegates 'on behalf of' the citizens is suspect, being in line with the neoliberal privatizing and depoliticizing aims (Harvey 2005).[20] And even innovative, anti-capitalist networks like the Open Source movement can be easily integrated into, or subjugated to, the logic of

19 This is hardly an exclusive feature of this literature. Vagueness about the role of 'civil society' in counteracting neoliberalism is widespread, with little signs of change in the most recent contributions (see e.g. Crouch 2011).

20 Of course, this does not amount to reducing NGOs to triggers of neoliberal hegemony, if anything because they respond in different ways to neoliberal forms of governance, for example as regards market-based environmental policies (see Blok 2011 and this volume).

capital (Suarez-Villa 2009). What the multitude theory fails to consider is how this self-constituted entity can become a revolutionary subject. The Gramscian argument remains valid: that counter-hegemonic forces should be able to take control over key institutions and processes by developing a new leadership at the same time economic, political and moral (Ransome 1992: 136). In this sense, the 'ecological dominance' that neoliberalism maintains over and through technoscience remains a crucial, unresolved issue.

Conclusion

We have argued that the ideational core of neoliberalism lies in a distinctive intensification of contrasting traits of the modern cultural heritage: an ontologically plastic world and a strong agency. This 'neorational' core turns post-structuralist accounts of the humanity-nature relationship upside down, translating a mutually constitutive interaction into a matter of appropriative control. The neoliberal entrepreneurial agent resembles a god, since the full pliancy of materiality to human designs leads to the depiction of agency in terms of an ultimately unconstrained, expansive will. Moreover, the oscillation of neoliberal discourses between description and prescription makes it particularly difficult to develop a critique. We have seen that technoscience is profoundly affected by neoliberal governmentality, as both a source and object of regulation, and that biotechnology is an elective terrain on which the remoulding of nature and humanity takes place. We have also seen that neoliberalism relies to a significant extent on the recodification of the political aspects of technoscience in terms of ethics or consensual dialogue, and that resisting neoliberalization presents major conceptual and practical challenges.

Compared with other readings which highlight the variety and complexity of neoliberalization and the countervailing tendencies detectable in the reciprocal constitution of science and neoliberalism (Brenner, Peck and Theodore 2010, Moore et al. 2011, Blok this volume, Hess this volume), our critique may appear one-sided. However, as indicated by the ambivalences of counter-science, ethics and participatory democracy, the varieties of and countervailing tendencies within neoliberalization have to be gauged against the latter's hegemonic character. This character is partly shared with other hegemonic constructions. Neoliberalism draws on a wide variety of culturally available objective thought forms, and seeks to naturalize a certain order by downplaying contrasting arguments and evidence. Moreover, its reliance on truth claims and on a close connection between science and politics belongs to the tradition of liberal democracy (Ezrahi 1990). Yet neoliberalism is also an 'intensified' form of hegemony in a twofold sense: it intensifies the traits of liberal governmentality and it intensifies hegemonic closure to counter-arguments and counter-conducts.

This feature surfaces in the difficulty of imagining a 'different' science. We concur with Brian Wynne's remark that 'for a considerable time society has been selectively directing not only technology but also (perhaps less directly so) scientific knowledge-inquiry and production – notwithstanding that lasting basic scientific understanding of nature has developed and accumulated alongside this more selectively applications-imagining, techno-scientific research activity' (Wynne 2011: xi). The enormous financial, organizational and professional efforts of the Human Genome Project, for example, make sense only by taking for granted a particular understanding (aligned with the neorational, entrepreneurial agency we have outlined) of what genes are and what we can, and may want, to do with them. And if the original Project has not fulfilled its eschatological expectations, research is still proceeding according to the same basic rationale.

From both a Foucauldian and a Gramscian perspective resistance is primarily a matter of practices, of counter-conducts. Our discussion suggests that such conducts should be targeted first and foremost on the hypertrophic neoliberal agency. The exit from the 'flexible paradise of neo-liberalism' (Lemke 2003: 64) is likely to entail a collectively practised rejection of commodification and appropriation, of makeability and enhancement, as driving values; a rejection grounded on recognition of a fundamental element of alterity of the biophysical and psychic world to human technoscientific agency.

Acknowledgment

We wish to thank Les Levidow for helpful comments and suggestions.

References

Abercrombie, N., Hill S. and Turner, B.S. 1980. *The Dominant Ideology Thesis*. London: Allen & Unwin.

Alanen, I. 1991. *Miten teoretisoida maatalouden pientuotantoa*. Jyväskylä: Jyväskylän yliopisto. Jyväskylä studies in education, psychology and social research 81.

Amoore, L. 2009. Algorithmic war: Everyday geographies of the war on terror. *Antipode*, 41(1), 49–69.

Anders, G. 2003. *L'uomo è antiquato*. Torino: Bollati Boringhieri, 2 Vols. (original edition *Die Antiquiertheit des Menschen*. München: C.H. Beck, 1956–1980).

Backstrand, K., Khan, J., Kronsell, A. and Lövbrand, E. 2010. (Eds.) *Environmental Politics and Deliberative Democracy: Examining the Promise of New Modes of Governance*. Chelthenham: Elgar.

Barnett, C. 2005. The consolations of 'neoliberalism'. *Geoforum*, 36(1), 7–12.

Barry, A. 2002. The anti-political economy. *Economy & Society*, 31(2), 268–84.

Bogner, A. and Menz, W. 2010. How politics deals with expert dissent: The case of ethics councils. *Science, Technology, & Human Values*, 35(6), 888–914.

Boyd, W., Prudham, S. and Schurman, R. 2001. Industrial dynamics and the problem of nature. *Society and Natural Resources*, 14, 555–70.

Braun, C., Herrmann, S.L., Könninger, S. and Moore, A. 2010. Ethical reflection must always be measured. *Science, Technology & Human Values*, 35(6), 839–64.

Brenner, N. Peck, J. and Theodore, N., 2010. Variegated neoliberalization: Geographies, modalities, pathways. *Global Networks*, 10(2), 182–222.

Brown, N. and Michael, M. 2003. A Sociology of expectations: Retrospecting prospects and prospecting retrospects. *Technology Analysis & Strategic Management*, 15(1), 3–18.

Brown, W. 2005. *Edgework. Critical Essays on Knowledge and Politics*. Princeton, NJ: Princeton University Press.

Burchell, G. 1996. Liberal government and techniques of the self, in *Foucault and Political Reason*, edited by A. Barry, T. Osborne and N. Rose. London: UCL Press, 19–36.

Calvert, J. 2007. Patenting genomic objects: Genes, genomes, function and information. *Science as Culture*, 16(2), 207–23.

Carolan, M. 2010a. The mutability of biotechnology patents: From unwieldy products of nature to independent 'object/s'. *Theory Culture & Society*, 27(1), 110–29.

Carolan, M. 2010b. *Decentering Biotechnologies. Assemblages Built and Assemblages Masked*. Farnham: Ashgate.

CEC. 1991. *Promoting the Competitive Environment for the Industrial Activities Based on Biotechnology Within the Community*. Brussels: Commission of the European Communities. SEC, 629 final.

Chiapello, E. 2003. Reconciling the two principal meanings of the notion of ideology: The example of the concept of the spirit of capitalism. *European Journal of Social Theory*, 6(2), 155–71.

Cooper, M. 2008. *Life as surplus. Biotechnology & Capitalism in the Neoliberal Era*. Seattle: University of Washington Press.

Cooper, M., 2010. Turbulent worlds. Financial markets and environmental crisis. *Theory, Culture & Society*, 27(2–3), 167–90.

Crouch, C. 2011. *The Strange Non-Death of Neoliberalism*. Cambridge: Polity Press.

Dean, M. 1999. *Governmentality*. London: Sage.

Drake, F. 2010. Protesting mobile phone masts: Risk, neoliberalism, and governmentality. *Science, Technology & Human Values*, 35(4), 522–48.

Ezrahi, Y. 1990. *The Descent of Icarus. Science and the Ttransformation of Contemporary Democracy*. Cambridge, MA: Harvard University Press.

Feher, M. 2009. Self-appreciation: Or, the aspirations of human capital. *Public Culture* 21(1), 21–41.

Felt, U. and Fochler, M. 2010. Machineries for making publics: Inscribing and describing publics in public engagement. *Minerva*, 48(3), 219–38.

Felt, U. and Wynne, B. 2007. (Eds.) *Taking European Knowledge Society Seriously*. Report for the European Commission. Luxembourg: Office for Official Publications of the European Communities. Available at: http://

ec.europa.eu/research/science-society/document_library/pdf_06/european-knowledge-society_en.pdf [accessed: 5 June 2012].

Foucault, M. 1980. *Power/Knowledge. Selected Interviews and Other Writings 1972–1977*. New York: Pantheon.

Foucault, M. 1982. How is power exercised?, in *Michel Foucault: Beyond Structuralism and Hermeneutics*, edited by H.L. Dreyfus and P. Rabinow. Brighton: Harvester Press, 216–26.

Foucault, M. 1991. Governmentality, in *The Foucault Effect: Studies in Governmentality*, edited by G. Burchell, C. Gordon, and P. Miller. London: Harvester Wheatsheaf, 87–104.

Foucault, M. 1997a. The ethics of the concern for the self as a practice of freedom, in *Essential Works of Foucault 1954–1984, Volume I: Ethics*, edited by P. Rabinow. New York: Penguin Books, 281–301.

Foucault, M. 1997b. What is critique?, in *The Politics of Truth*, edited by S. Lotringer. Los Angeles: Semiotext(e), 41–82.

Foucault, M. 2000. The subject and power, in *Essential Works of Foucault 1954–1984, Volume III: Power*, edited by J. Faubion. New York: Penguin Books, 326–48.

Foucault, M. 2003. '*Society must be defended'. Lectures at the Collège de France 1975–1976*. New York: Picador.

Foucault, M. 2008. *The Birth of Biopolitics*. London: Palgrave MacMillan.

Freudenburg, W., Gramling, R., and Davidson, D. 2008. Scientific certainty argumentation methods (SCAMs): Science and the politics of doubt. *Sociological Inquiry*, 78(1), 2–38.

Frickel, S., Gibbon, S., Howard, J., Kempner, J., Ottinger, G. and Hess, D. 2010. Undone science: Charting social movement and civil society challenges to research agenda settings. *Science, Technology, and Human Value*s, 35(4), 444–73.

Goodin R. 2000. Democratic deliberation within. *Philosophy & Public Affairs*, 29(1), 79–107.

Goven, J. 2006. Dialogue, governance and biotechnology: Acknowledging the context of the conversation. *Integrated Assessment Journal*, 6(2), 99–116.

Gramsci, A. 1978. *Selections from the Prison Notebooks*, edited and translated by Q. Hoare and G. Nowell Smith. London: Lawrence & Wishart.

Hall, S. 1983. The Problem of Ideology – Marxism without guarantees, in *Marx – A Hundred Years On*, edited by B. Matthews. London: Lawrence & Wishart.

Hardt, M. and Negri, A. 2004. *Multitude: War and Democracy in the Age of Empire*. New York: Penguin Press.

Harvey, D. 1996. *Justice, Nature and The Geography of Difference*. Oxford: Blackwell.

Harvey, D. 2003. *The New Imperialism*. Oxford: Oxford University Press.

Harvey, D. 2005. *A Short History of Neoliberalism*. Oxford: Oxford University Press.

Harvey, D. 2006. Neoliberalism as creative destruction. *Annals of the American Academy of Political and Social Science*, 610, 22–44.

Hayden. C. 2007. Taking as giving: Bioscience, exchange, and the politics of benefit-sharing. *Social Studies of Science*, 37(5), 729–58.

Heynen, N., McCarthy, J., Prudham, S. and Robbins, P. 2007. Introduction: False promises, in *Neoliberal Environments: False Promises and Unnatural Consequences*, edited by N. Heynen, J. McCarthy, S. Prudham and P. Robbins. London: Routledge, 1–21.

Hindess, B. 1996. *Discourses of Power. From Hobbes to Foucault*. Oxford: Blackwell.

Hunt, A. 2004. Getting Marx and Foucault into bed together. *Journal of Law and Society*, 31(4), 592–609.

Irwin, A. 2006. The politics of talk: Coming to terms with the 'new' scientific governance. *Social Studies of Science*, 36(2), 299–320.

Jessop, B. 2002. Liberalism, neoliberalism, and urban governance: A state-theoretical perspective. *Antipode*, 34 (3), 452–72.

Jessop, B. 2007. From micro-powers to governmentality: Foucault's work on statehood, state formation, statecraft and state power. *Political Geography*, 26(1), 34–40.

Jessop, B. 2010. From hegemony to crisis? The continuing ecological dominance of neoliberalism, in *The Rise and Fall of Neoliberalism*, edited by K. Birch and V. Mykhnenko. London: Zed Books, 171–87.

Kay, J. 1999. In the beginning was the word: The genetic code and the book of life, in *The Science Studies Reader*, edited by M. Biagioli. New York: Routledge, 224–33.

Kinchy, A., Kleinman, D. and Autry, R. 2008. Against free markets, against science? Regulating the socio-economic effects of biotechnology. *Rural Sociology*, 73(2), 147–79.

Laclau, E. and Mouffe, C. 1985. *Hegemony and Socialist Strategy. Towards a Radical Democratic Politics*. London: Verso.

Larner, W. 2003. Neoliberalism? *Environment and Plannning D*, 21(5), 509–12.

Latour B. 1993. *We Have Never Been Modern*. Cambridge, MA: Harvard University Press.

Lazzarato, M. 2009. Neoliberalism in action: Inequality, insecurity and the reconstitution of the social. *Theory, Culture & Society*, 26(6), 109–33.

Lemke, T. 2003. Foucault, governmentality and critique. *Rethinking Marxism*, 14(3), 49–64.

Lemke, T. 2004. 'Disposition and determinism – genetic diagnostics in risk society'. *Sociological Review*, 52(4), 550–66.

Lemke, T. 2007. An indigestible meal? Foucault, governmentality and state theory. *Distinktion: Scandinavian Journal of Social Theory*, 8(2), 43–64.

Levidow, L. 2007. European public participation as risk governance: Enhancing democratic accountability for agbiotech policy? *East Asian Science Technology and Society*, 1, 19–51.

MacLeavy, J. 2008. Managing diversity? 'Community cohesion' and its limits in neoliberal urban policy. *Geography Compass*, 2(2), 538–58.

Mattelart, A. 2010. *The Globalization of Surveillance. The Origin of the Securitarian Order*. Cambridge: Polity Press.
Mirowski, P. and Plehwe, D.2009. (Eds.) *The Road From Mont Pelerin: The Making of the Neoliberal Thought Collective.* Cambridge, MA: Harvard University Press.
Mitchell, R. and Waldby, C. 2010. National biobanks: Clinical labor, risk production, and the creation of biovalue. *Science, Technology, & Human Values*, 35(3), 330–55.
Mol, A. and Law, J. 2006. Complexities: An introduction, in *Complexities. Social Studies of Knowledge Practices*, edited by J. Law and A. Mol. Durham, NC: Duke University Press, 1–22.
Moore, K., Kleinman, D., Hess, D and Frickel, S. 2011. Science and neoliberal globalization: A political sociological approach. *Theory and Society*, 40(5), 505–32.
Mouffe, C. 2005. *On the Political*. London: Routledge.
Myers, E. 2008. Resisting Foucauldian ethics: Associative politics and the limits of the care of the self. *Contemporary Political Theory*, 7, 125–46.
Nealon, J. 2008. *Foucault Beyond Foucault. Power and its Intensification since 1984*. Stanford: Stanford University Press.
O'Malley, P. 2004. *Risk, Uncertainty and Governance*. London: Glasshouse.
O'Malley, P. 2008. Governmentality and risk, in *Social Theories of Risk and Uncertainty*, edited by J. Zinn. London: Blackwell, 52–75.
Ong, A. 2006. *Neoliberalism as Exception*. Durham: Duke University Press.
Ottinger, G. 2010. Buckets of resistance: Standards and the effectiveness of citizen science. *Science, Technology, & Human Values*, 35(2), 244–70.
Pellizzoni, L. 2010. Risk and responsibility in a manufactured world. *Science and Engineering Ethics*, 16(3), 463–78.
Pellizzoni, L. 2011a. Governing through disorder: Neoliberal environmental governance and social theory. *Global Environmental Change*, 21(3), 795–803.
Pellizzoni, L. 2011b. The politics of facts: Local environmental conflicts and expertise. *Environmental Politics*, 20(6), 765–85.
Plehwe, D., Walpen, B. and Neunhöffer, G. 2005. (Eds.) *Neoliberal Hegemony. A Global Critique*. London: Routledge.
Polanyi, K., 1944. *The Great Transformation.* Boston: Beacon Press.
Rabinow, P. 1996. Artificiality and enlightenment: From sociobiology to biosociality, in *Essays in the Anthropology of Reason*. Princeton, NJ: Princeton University Press, 91–111.
Ransome, P. 1992. *Antonio Gramsci. A New Introduction*. Hertfordshire: Harvester Wheatsheaf.
Read, J. 2009. A Genealogy of homo-economicus: Neoliberalism and the production of subjectivity. *Foucault Studies*, 6, 25–36.
Robinson, W.I. 2004. *A Theory of Global Capitalism. Production, Class and State in a Transnational World.* Baltimore: Johns Hopkins University Press.

Roco, M. and Bainbridge, W. 2002. (Eds.) *Converging Technologies for Improving Human Performance*. Arlington, VI: National Science Foundation. Available at: http://www.wtec.org/ConvergingTechnologies/1/NBIC_report.pdf [accessed: 20 December 2011].

Rose, N. 1996. Governing 'advanced' liberal democracies, in *Foucault and Political Reason*, edited by A. Barry, T. Osborne and N. Rose. London: UCL Press, 37–64.

Rose, N. and Novas, C. 2005. Biological citizenship, in *Global Assemblages*, edited by A. Ong and S. Collier. London: Blackwell, 439–63.

Rose, N. 2007. *The Politics of Life Itself*. Princeton, NJ: Princeton University Press.

Salter, B. and Salter, C. 2007. Bioethics and the global moral economy: The cultural politics of human embryonic stem cell science. *Science, Technology, & Human Values*, 32(5), 554–81.

Sayer, D. 1979. *Marx's Method. Ideology, Science and Critique in Capital*. Sussex: Harvester Press.

Suarez-Villa, L. 2009. *Technocapitalism*. Philadelphia: Temple University Press.

Szerszynski, B., Heim, W., and Waterton, C. 2003. (Eds.) *Nature Performed: Environment, Culture and Performance*. Oxford: Blackwell.

Tallacchini, M. 2009. Governing by values. EU ethics: Soft tool, hard effects. *Minerva*, 47(3), 281–306.

Tazzioli, M. 2011. *Politiche della verità. Michel Foucault e il neoliberalismo*. Verona: Ombre Corte.

Thaler, R. and Sunstein, C. 2008. *Nudge. Improving Decisions about Health, Wealth, and Happiness*. New Haven, CT: Yale University Press.

Tickell, A. and Peck, J. 2003. Making global rules: globalization or neoliberalization?, in *Remaking the Global Economy*, edited by J. Peck and H.W. Yeung. London: Sage, 163–81.

Urbinati, N. 2010. Unpolitical democracy. *Political Theory*, 38(1), 65–92.

Virno, P. 2004. *A Grammar of the Multitude. For an Analysis of Contemporary Forms of Life*. Los Angeles, CA: Semiotext(e).

Virno, P. 2009. Natural-historical diagrams: The 'new global' movement and the biological invariant. *Cosmos and History*, 5(1), 92–104.

Walker, G., Cass, N., Burningham, K. and Barnett, J. 2010. Renewable energy and socio-technical change: Imagined subjectivities of 'the public' and their implications. *Environment and Planning A*, 42(4), 931–47.

Ward, K. and England, K. 2007. Introduction: Reading neoliberalization, in *Neoliberalization: States, Networks, Peoples*, edited by K. England and K. Ward. London: Blackwell, 1–22.

Wynne, B. 2001. Creating public alienation: Expert discourses of risk and ethics on GMOs. *Science as Culture*, 10, 1–40.

Wynne, B. 2005. Reflexing complexity. *Theory, Culture & Society*, 22(5), 67–94.

Wynne, B. 2011. Foreword, in A. Feenberg. *Between Reason and Experience. Essays in Technology and Modernity*. Cambridge, MA: MIT Press, viii–xv.

Ylönen, M. 2011. *Saastumisen kontrollin ideologia. Vesien saastumisen ja vesirikosten kontrollin ideologia Suomessa vuosina 1960–2000*. [*Ideology of pollution control. Social control of water pollution and pollution crimes in Finland from the 1960s until the year 2000*]. Jyväskylä Studies in Education, Psychology and Social Research, 424. Jyväskylä: Jyväskylä University Press.

Chapter 3

Neoliberalism and ICTs: Late capitalism and technoscience in cultural perspective

Alessandro Gandini

Introduction

This chapter explores the role of ICTs in the relationships between neoliberalism on one side and technoscientific evolution on the other, in cultural perspective, within the late capitalist setting (Mandel 1975, Jameson 1984). The hypothesis this work aims to illustrate is that Information and Communication Technologies (simply ICTs from now on) and, in a larger picture, a culturalist approach entailing media in society, are crucial in order to understand the relationship between neoliberalism and technoscience in this context. ICTs are here conceived as an example of technoscience in everyday life, functioning as instruments which trace particular sorts of society (Michael 2006). I believe that, in this regard, we need to consider society as a common ground or milieu in which neoliberalism and technoscience have encountered and intertwined due to the increasingly social and cultural nature of late capitalism during the last forty or fifty years. Within this frame, as we will see, ICTs occupy a highly relevant position.

The following lines will provide a theoretical argumentation in which I try to show that we cannot understand the relationships between neoliberalism and technoscience in late capitalism without taking into account ICTs and media culture, over which neoliberalism operated as a political-economic vision. Let me point out right from the start that I am not supporting here any naively deterministic approach which argues that information technologies have originated an anthropological, sociological or cultural transformation in society. My position in this sense is that, from a cultural perspective, ICTs stand as the condition of possibility of the emergence of neoliberalism and, as we will see, even for its potential overcoming. We could say that the evolution of ICTs and neoliberalism effectively developed in the form of two separate trajectories which, although inextricably linked, slowly but consistently diverged to the extent that the digital development of ICTs is now potentially posing the foundations for the overcoming of neoliberalism itself.

The setting: Neoliberalism and late capitalism in cultural perspective

Neoliberalism

To begin with, let me outline what is conceived here for neoliberalism. In the book *A Brief History of Neoliberalism* David Harvey (2005) shows the historical processes which gave rise to the neoliberal 'ideology'. According to Harvey, and I adhere to this claim, neoliberalism can fairly be called an ideology since it has effectively established as the socio-economic leading vision for a large number of Western capitalist countries in the latest part of the last century, dramatically alternative to the socialist-communist model, to then soon become a somewhat hegemonic political-economic approach to the extent that, in the same West, it has come to coincide in practice with capitalism itself. Neoliberalism in Harvey's description is connoted by the inspiring belief that a free and almost completely deregulated market with little or no state intervention, characterized by a strong emphasis on individualism and private property, would enhance entrepreneurship and individual success in business, leading to a generally wealthier society.

However, this, according to Harvey, has not happened; actually, he argues that the main consequences of the diffusion of neoliberalism have been on the one hand the destruction of the institutional framework due to the limited role of the State, and on the other hand a strengthening of the division of labour, leading to a non-homogeneous welfare. Harvey traces the roots of neoliberalism back to the late 1970s, conceiving it as the outcome of the 1973 oil crisis which at that time consistently affected capital accumulation. The solution to that crisis was found, in the West, in a consistent push towards powerful market-centred, state-free, financial-based neoliberal policies in the early 1980s, whose main political protagonists in the West go from Margaret Thatcher in the UK and Ronald Reagan in the US in the early 1980s, up to George W. Bush in the first decade of the 2000s. The outcome of neoliberal policies, in Harvey's view, has been the emergence of a model in which 'ethics' coincided with the idolization of the market, and profit became the only end to pursue by any means.

This process can be framed according to Giovanni Arrighi's *The Long Twentieth Century* (first edition 1994), in which the author builds up a model to outline the development of capitalism through 'systemic accumulation cycles' from the Middle Ages to the globalized economy (Arrighi 1994). Each cycle, he argues, is characterized by the presence of a country or state which plays a leading hegemonic role, and has a sort of wave-like trend for which, in the first part of the 'curve', capital is accumulated through territorial expansion and 'material' accumulation until a 'signal crisis' takes place; the outcome of this crisis is the opening of new spaces of capital accumulation through financial investments. Later, when financial markets become again saturated, a new powerful financial crisis, that he calls 'terminal crisis', ends up the whole cycle, allowing another country to take over during the 'tail' of the ending cycle and become the new leader for a new process of accumulation to begin. I am here drawing from

Arrighi the idea that neoliberalism coincides with the second part of the more recent 'accumulation cycle', that he calls 'the long twentieth century', and which is connoted by the predominant position of the US in world economy and politics. In Arrighi's account, the 'American cycle' is characterized by a 'material' phase characterized by World Wars I and II up to the Vietnam campaign and which ended in the already mentioned 1970s oil downturn, that Arrighi interprets as the 'signal crisis'. Hence, starting from the late 1970s and early 1980s this cycle has developed through financial accumulation. In Arrighi's account, this is the age of neoliberalism, conceived as the hegemonic ideology of a financial-based period of accumulation and economic growth which lasted, we would add, until 2007, when another crisis hit the markets. If we stick to Arrighi's model, this is likely to be the terminal crisis for the last accumulation cycle.

We will come back to this issue later in this work; for now, I am taking Arrighi's account merely as a map to outline how cycles of accumulation work and, more peculiarly, to frame neoliberalism in a socio-historical dynamic. The emphasis on 'crises' is indeed crucial in this context, since it is from one of these crises (mid-1970s) that a political-economic vision pushing for market deregulation emerged as a possible solution towards economic reprise. This is what effectively happened, since neoliberal policies actually resulted in order to drag out Western economies from the 'signal crisis' and ignite a period of growth. However, there is more to the picture than meets the eye. Neoliberalism in fact operated within a complex and variously articulated scenario where ICTs and media culture in broader terms play a crucial role.

Late capitalism in cultural perspective

In a seminal article appeared on the *New Left Review* in 1984, Fredric Jameson argues that 'postmodernism' is the Cultural Dominant of the final part of the twentieth century, representing the overcoming of 'modernism' in terms of a symbolic 'fall' of the grand, master narratives which characterized most of the twentieth century. More specifically, postmodernism in Jameson's view is 'the cultural logic of late capitalism', since culture and aesthetics have been incorporated into the actual mode of production. Jameson believes there is an inextricable connection between postmodernism and the nature of what he calls 'multinational capitalism' in the early 1980s.

This consists in two main elements. On the one hand, postmodernism bears a general sense of recognition and enthusiastic acceptance of the presence and relevance of mass media in society, to the extent that monopoly capitalism has effectively swallowed up the previous notion of 'public sphere' into that of a 'mass-mediatized' public opinion. On the other hand, it configures the emergence of a so-called 'post-Fordist' mode of production (Jameson 1984), constituted of non-standardized and socially-tailored production characterized by a decline of manufacture and a growing importance of knowledge-based service and information industries, with a process that Harvey (1989) calls 'flexible accumulation'. ICTs here represent the social and cultural milieu into which this process of growing importance of cultural production in social life emerged.

Neoliberalism, on the other hand, represented a political-economic vision which operated in an economic setting with increasingly social traits. More precisely, what I am arguing here is that late capitalist society is connoted by what Guy Debord calls 'society of the spectacle', in other words a society 'which prefers the sign to the thing signified, the copy to the original, representation to reality, the appearance to the essence' (Debord 1967: 11). In the 'society of the spectacle' everything is commodified and commodities are '*all* that there is to see; the world we see is the world of commodity' (Debord 1967: 29). Media age configures itself as the historical moment in which the commodity completes its 'colonization of social life' (Debord 1967: 29) to the extent that the spectacle becomes a social relationship between people mediated by images (Debord 1967) from which *sign value* is extracted (Baudrillard 1981). In late capitalism, in other words, commodities flow in the form of images and cultural objects (Appadurai 1986) across information and communication technologies.

The element connecting all these instances to neoliberalism is brand. Celia Lury defines brand as an 'object' whose nature involves images, processes, products and relations between products. She argues that brand works as an interface of communication between producers and consumers, claiming that brand is a media object and a medium of communication, functioning as a mark of social identities flowing in and out of the context of ICTs (Lury 2004). Drawing from Lury's account we can say that brand is the 'objectified' relational node where commodity and *sign-value* (Baudrillard 1981) intertwine in the embodiment of the commodification of the image, in other words the place where the image of the commodity coincides with the commodity in itself. Brand therefore stands as the linking element which connects neoliberalism, post-Fordism and ICTs since symbolic objects in global popular culture circulating in and out of media have increasingly established as managerial tools and intangible assets able to attract large financial investments in deregulated markets. Intangible and immaterial assets have, in other words, become the fundamental element to extract value in late capitalist, neoliberal economy to the extent that, with the emergence of digital and social media, the production of value has increasingly become social and now even offers economic visions and models which are potentially alternative to neoliberalism (Arvidsson 2006, 2011).

To conclude, we can say that neoliberalism as a vision enhancing individual entrepreneurship, private property and social success through business (Harvey 2005) mirrored the evolution of ICTs in the form of broadcast media enhancing personal success and individual social recognition through consumption of symbolic goods in the 'society of the spectacle' (Debord 1967). It is in ICTs and in the performative context that they were able to fuel, that neoliberalism found a fertile ground to develop successfully; and this is why, from a cultural perspective, ICTs in the form of broadcast media stand as the condition of possibility of the emergence of neoliberalism. This model remained successful until economic growth has ignited consumerism; the 2007 crisis, indeed, came and affected also consumerism, which is crucial to the success of neoliberal policies, exactly at the same time when digital media were showing potentially alternative models alongside the social production and social creation of value which is peculiar to post-Fordism – I will talk about this later on.

Here, indeed, we have the diverging trajectories: the success of neoliberalism was, at least in part, due to a vision of life based on consumption brought along by broadcast media, which has hidden the contradiction of an intrinsically individualistic socio-economic model exerted upon an increasingly social mode of production and value creation – that is post-Fordism. The 2007 crisis unveiled the contradiction when new tools, networked digital media, allowed the process of evolution of value creation towards social production to fully develop. The reason for this claim lies on the 'opposite shore', that is in the nature of post-Fordism and the evolution of labour in the period considered.

ICTs and post-Fordism

The multitude

According to Paolo Virno, the emergence and evolution of post-Fordism involves a reconsideration of the dialectic between 'the people' and 'the multitude', a philosophical question started with Spinoza in the seventeenth century. Virno, in the work *A Grammar of the Multitude* (Virno 2004), argues that, with the emergence of interactive media and following the development of the relationships between politics and labour during the past twenty years, we are now confronted to a social landscape whose main actor is no longer 'the people', rather the Spinozian 'multitude', as outlined also by Michael Hardt and Antonio Negri (2000, 2004).

Virno claims that the notion of multitude finds itself inserted into a more complex analysis involving perception and subjectivity. He argues that the 1700s debate between Thomas Hobbes and Baruch Spinoza on the concepts of people and multitude saw Hobbes prevail: 'the people' became the main subject of industrial society as a product of the State into the context of industrialization. In other words Virno claims that, at the dawn of the 'modern' industrial era, 'the people' as a social agent with characteristics of unity and 'oneness', especially at the level of subjectivity, became predominant and used to reflect in the Fordist setting the idea of 'class' and especially 'working class', that is to say a collective political agent with social consciousness (Virno 2004).

The notion of class, Virno argues, was undoubtedly effective in Fordism. However, with the emergence and the diffusion of the mass media, the situation began to change. Virno claims that, in the passage to '*post*-Fordism', the Spinozian concept of 'multitude', intended as 'a plurality which persists as such in the public scene […] without converging into a One' (Virno 2004: 21) returned to manifest itself. The reason of this statement lies specifically in the acceptance of a progressive mediatization of society; Virno argues that mass media realized a blending of the private into the public which led to a modification at the levels of perception and subjectivity, to the extent that a 'newly' singular, heterogeneous social agent reappeared on the social scene: the member of the multitude. This process, in Virno's opinion, became visible only after the fall of the Berlin Wall in 1989; the multitude re-emerged when the ideological 'enemy' is beaten, and there is no longer a 'hostile outside' especially at the political level, that is the place where 'the people' used to express its subjectivity as 'class' in full.

The absence of a visible, explicitly perceived element of 'struggle', mingled with the impact of the mass media at a social and cultural level, according to Virno, allowed the re-emergence of 'the multitude' as the main social agent, displacing 'the people'. In this sense, Virno (2004: 21–47) defines the multitude as a middle region, a 'grey zone' between the individual and the collective. In fact, whereas 'the people' was an eminently collective subject whose aim was to converge towards an ideal unity or 'collective subjectivity', on the other hand 'the multitude', thanks to the already mentioned elements, presents a remarkable individual and heterogeneous trait in which unity is 'the premise' for individuality.

In order to explain his theory Virno draws from Gilbert Simondon's concept of individuation, developed in the work *L'individuation pshychique et collective* (Simondon 1989), to describe the multitude as a plurality of individuals whose common shared element lies in the Simondonian 'pre-individual', in other words at the cognitive level of language and perception. Class consciousness, in Virno's opinion, is no longer the One as universal subjectivity to aim at. Conversely, the One, in the multitude, is a pre-individual, universal element pre-existing the multitude's individual, which comes 'after' in the form of a social entity which finds itself individuated no longer in a promise of universality, but rather in a process of singular self-individuation coming from a shared – cognitive, linguistic – universal. To define this common shared ground Virno draws from Marx (1973) the notion of General Intellect (GI), whose main feature lies therefore at a cognitive level and presents itself, in Simondon's lexicon, as the contemporary mode of being of a social agent which individuates himself in a process of differentiation 'ex-post'. Before this, with the 'people' we had 'differentiation' as the premise and 'unity' as the promised outcome – la *volonté générale*. Now, 'differentiation' becomes the outcome, and 'unity' the premise, which is displaced from the socio-political ground and lies at a pre-individual level.

As a result, the dialectic between subjectivity and universality in the notion of class appears to be subverted, although not completely dismissed. In fact, Virno clarifies that the concept of GI does not imply in any sense the end of labour class, rather a modification in its nature. Virno explains that GI means that the modes of organization of subjectivity no longer assume the traits of an organized unity as it was in relation to the people. He argues that the 'public sphere', thanks to the mass media, has been displaced from the site of labour and politics, in which it was previously located as class consciousness and substantial community, to become a 'non-governmental public sphere'. The *res publica* is no longer the common ground of the public sphere: all participants are bound together by a publicness located into the media which stands, broadly speaking, as a heterogeneous set of individualities. Virno (2004: 40–42) talks explicitly of a 'publicness without a public sphere'.

The notion of 'multitude' has been object of quite extensive criticism in the literature. For instance David Harvey (2003), discussing the notion of multitude outlined by Hardt and Negri (2000) emphasizes the mechanical and passive trait of the multitude, which he calls 'amorphous'. Harvey argues that this concept needs to be put at work in a more cohesive form in terms of 'connectivity within struggle', so that it can effectively become a political entity in what he conceives as the opposition to 'imperialism' (Harvey 2003). Similarly, Chantal Mouffe (2008) underlines the fact that the notion of struggle in the 'Operaist'[1] account is connoted by a resistant and oppositional trait without being fruitfully antagonist – and therefore truly alternative – to the political-economic system they criticize. Mouffe argues that what is at stake in the critique to the multitude is the lack of an attempt of articulation of this missing collective identity, which in the form of the GI remains a purely immanent potentiality. Mouffe stresses the fact that Virno's (2004) and also Hardt and Negri's (2000, 2004) accounts crucially lack political articulation as antagonist to the dominant hegemony; this construction, in her view, is needed in order to make their critique effective. This political articulation, although with differences, is to be found for both Harvey (2003) and Mouffe (2008) in connectivity, not in self-enclosed communities or social groups (Harvey 2003) but rather through linking social movements to political parties and trade unions (Mouffe 2008).

To conclude, we could arguably claim that the 1989 Berlin events can also be read as the historic success of neoliberal capitalism over socialist-communist ideology; however, quite curiously, Virno seldom talks of neoliberalism in explicit terms. Indeed, we could say that the visible outcome of that victory can be found in a change in the nature of the working class resulting in a fragmentation of workers' subjectivity and in a growing relevance of the so-called 'knowledge workers'. In order to enquire about this modification and to justify this claim it is therefore needed to focus more specifically on the nature of labour in late capitalism.

Immaterial labour and cognitive capitalism

The Italian, France-based sociologist Maurizio Lazzarato labels as 'immaterial labour' (Lazzarato 1996b, 1997) the notion that allows to re-define the whole concept of commodity at the end of the twentieth century. Immaterial labour, in

1 The term 'Operaismo' ('Workerism') refers to a leftist, post-Marxist stream of research grown in Italy in the 1960s and developed by thinkers such as Antonio Negri and Mario Tronti and, more recently, Paolo Virno and Maurizio Lazzarato. For further information see Mezzadra (2000) and Wright (2002).

Lazzarato's words, is labour 'that produces the informational and cultural content of a commodity' (Lazzarato 1996b: 133); with this statement Lazzarato intends to claim that the post-Fordist commodity has changed its 'material' nature to become an eminently 'informational' and 'cultural' object. The term 'informational' refers to the 'immateriality' of the contemporary commodity, which is increasingly made of abstract data and incorporeal entities, whereas 'cultural' means that the commodity is no longer a 'product' resulting from industrial labour, rather the outcome of activities which are not simply definable as 'work'. Following Virno and his 'Simondonian' theorizations, Lazzarato argues that the working class labour has now become an abstract activity involving, at a deeper level, the application of subjectivity in forms which are immediately collective – that is, the GI and its cognitive traits (Lazzarato 1996b: 133–6).

More so, the notion of 'immaterial labour' allows Lazzarato to open up a broader discussion claiming that the post-Fordist regime in its latest developments forces a reconsideration of the main categories of production and consumption. In Lazzarato's account the Fordist corporation which generates commodities has historically evolved in the 'post-Fordist corporation' which no longer merely generates the commodity-object but also the world in which the commodity will have to exist in. Lazzarato claims, similarly to Debord (1967), that the process just displayed is mostly due to the mass media, which are crucial in constructing the universe of 'immaterial' values around the commodity object (Lazzarato 1996a, 1996b).

The mechanism of consumption, in Lazzarato's view, no longer means the mere act of buying or carrying out a service according to a desire or need, as traditional economic theory claims; rather it displays a process of universe-creation which is socially, culturally and cognitively constructed. It is exactly what 'brands' do, as we have seen earlier (Lury 2004, Arvidsson 2006), leading to the creation of 'brandscapes' (Klingmann 2007). Advertising and brands have become, according to Lazzarato, a 'spiritual' process to design a 'social prompt' in which the commodity goes to 'perform'. This configures, in Lazzarato's view, a process of subjectivation that reaches the consumer's cognitive traits as a productive and performative refrain (Lazzarato 1996b). This process of universe-creation is indeed, effectively, a mechanism of value creation. To Lazzarato, 'the post-industrial enterprise and economy' consists in a mechanism of 'production and consumption of information' (Lazzarato 1996b) where both use-value and exchange-value are to be conceived in increasingly cognitive terms. As a result, we are shifting from a 'model of consumption' to a 'model of communication', in which communication stands as a new form of Marxian superstructural interface between production and consumption. In other words, 'immaterial labour' works as a communication process in which the commodity per se is not destroyed in the act of consumption, but finds itself enlarged as the 'creator' of the consumer's environment and world. Particularly, Lazzarato argues in explicit analogy with Virno that 'immaterial labour' is a process of 'production of communication' tending to immateriality through aesthetics (Lazzarato 1996b).

A few years later, Lazzarato refines his claims asserting that capitalism in its latest, post-Fordist form appears no longer to be 'merely' a mode of production, but rather a 'mode of production of modes of production' in which communication industry takes shape as the industry of the means of production. This industry does not consist in producing machinery, but rather in the production of those 'immaterial', cognitive competencies which appear now inseparable from living labour (Lazzarato 2004). This dynamic has been called 'cognitive capitalism' (Moulier-Boutang 2007) or 'bio-capitalism', a form of capitalism which aims to extract value from every aspect of life (Fumagalli 2007). More specifically, 'cognitive capitalism' is defined by Yann Moulier-Boutang as the third phase of capitalism, following the mercantile and the industrial period. The rationale of the notion of 'cognitive capitalism', which explicitly draws from Lazzarato's concept of 'immaterial labour', lies precisely in its 'immaterial' nature, a mainly individual and heterogeneous process which favours cognitive skills over technicality (Moulier-Boutang 2007). Again, according to this theoretical approach, the development of information and communication technologies led to a mutation in the nature of labour to the extent that, as Antonio Negri also writes, 'the whole labour process is tending towards immaterial labour' (Negri 2008: 75).

In other words, the milieu in which this cognitive process found a fertile terrain is the digital environment and what Lazzarato (1996a) calls 'numerical networks'. Despite being an abused expression, numerical networks effectively appear to be something very different from the previous media, in many ways. Firstly, they provide an 'infrastructure' through which the meta-communicative perspective could finally appear, so that – as already said – Lazzarato defines this communication-based post-Fordism 'a mode of production of modes of production' (Lazzarato 2004). Secondly and more importantly, after an initial age in which interactive and digital media used to reproduce the static structure of signification peculiar to the mass media, quite soon this infrastructure effectively became a networked milieu for social interaction and economic activity.

The commodity-object described above, in the context of digital and social media, was finally able to express its post-Fordist 'social potential' in full, to the extent that it now appears to have become something quite distant from the neoliberal commodity in the society of the spectacle. The neoliberal model of consumption now seems to be under threat by 'different' economic approaches at the rationale of which we have 'commons' and processes of 'sharing' that imply less or no private property, a renaissance of use value over exchange value (Arvidsson 2009) and, crucially, no more room for application of those neoliberal values (Harvey 2005) which, as we have seen, used to find especially in broadcast media a prolific terrain for development.

Knowledge work, creativity, technocapitalism: The aftermath of an era?

As we have seen, Virno argues that cognitive, pre-individual elements have increasingly come to the forefront vis-à-vis material work skills. Technical requirements and competences in late capitalism have become secondary in relation to knowledge, language, thought, self-reflection, communication and adaptability to work in teams. Virno concludes that we are now confronted with a labour class whose relational trait is no longer outlined in a base-superstructure logic at a class level, but appears to be one of a 'personal dependency', where the worker appears to be reduced to his/her human cognitive basics. This, according to Virno, provides evidence of the failure, in the present context, of the traditional Marxian tripartition Labour/Political Action/Intellect, which now appears to be hybridized together in a 'new' frame that he calls 'virtuosity'. Language and labour, in Virno's view, become part of a self-referential process that he labels as 'virtuoso', by this term meaning 'without end product' (Virno 2004: 47–94).

With the term 'virtuosity', therefore, Virno describes the a-productive, self-referential activity of the post-Fordist *knowledge worker*, conceived as an emerging middle class of workers connoted by freedom and self-organization (Drucker 1942, Bell 1973, Reich 1992, Pratt 2008) whose 'labour' appears to be sharing characteristics with the 'performance' and the 'score' of a performing artist. If we enlarge this picture, we can argue that the neoliberal ideology of profit was 'contaminated' by post-Fordism to the extent that intellectual skills, creativity and knowledge increasingly played a crucial role in a dynamic that Enzo Rullani (1998), in a slightly different approach from Lazzarato's and Virno's neo-Marxism, calls 'knowledge capitalism'. Rullani argues that the inter-dependent relationship between capitalism and technology is in fact nothing new: the history of capitalism is actually built upon the application of technoscientific development to the productive machine.

Compared with Lazzarato's 'Deleuzian' approach that, as we have seen, intends numerical networks as capable of actualizing a series of 'virtual' possibilities previously unfulfilled, Rullani distances himself claiming for an evolution of capitalism and post-Fordism alongside the evolution of networks in the same way as any other technology did in the past, in the sense of a merging and hybridization between markets and networks into a structure which ideally serves to boost value production in economic terms, without seeing any specificity in contemporary numerical networks (Rullani 1998). Drawing from Daniel Bell's analysis on post-industrial society (Bell 1973), Nico Stehr (1994) has argued in a similar way about what he calls 'knowledge societies', claiming that the latter is no longer organized through individuals and machines with the purpose of producing commodities *per se*; rather, it is centred on knowledge, whose productivity relies inevitably on people, who are the real actors in a scenario in which immaterial production has increased and come to the forefront alongside traditional production.

The outcome of this increasing importance of knowledge in late capitalism, from a neoliberal perspective, has been the emergence in the last decades of what has been called *creative* or *cultural industries*, which can be broadly defined as industries in which economic value is generated through mechanisms of valorization of knowledge, innovation and creativity (Hesmondhalgh 2002, Florida 2002). Creative industries appear to be, to a certain extent, a consequence of the application of neoliberal values over the knowledge-based setting, as the spreading of self-employment and freelancers seems to show (Christopherson 2008). Labour, from this perspective, becomes one among many commodities; a commodity which is produced and socially consumed in 'reputational' terms, according to a process which I would call 'labour consumption', where self-branding operates with the aim of achieving symbolic profit through social success thanks to the quality of the job and the 'scene' it actually allows the creative worker to participate in. In the creative industries, in fact, knowledge workers have been also defined as a 'creative class' (Florida 2002), by this term meaning an undistinguished ensemble of young workers employed in widely different jobs, from journalism to advertising and marketing, who share a common ethos for creativity, innovation and individualism. Considering the existence of a 'creative class' in the terms conceived by Florida, however, is indeed worthy of some criticism since this appears to be a mere list of professional figures rather than a real class; the traits they share are too general, more related to their lifestyle rather than to their social status, so that this appears to be too broad a classification. More so, and crucially for this work, the classification provided by Florida is connoted by a generally enthusiastic mood which indeed, after several years from his book, seems largely outdated. Creative workers are in fact usually connoted by a subtle ambivalence for which they can profit from a 'bright side' made of a job and a lifestyle which allows to enjoy freedom and self-organization, but on the other hand quite often incur into experiencing a 'dark side' made of low-paid or unpaid jobs, alienation, precariousness and the backlash of an increasing confusion between leisure and work time (Lazzarato 1997).

In a larger perspective, indeed, what is at stake here is what Luis Suarez-Villa (2009) calls 'technocapitalism'. In Suarez-Villa's model, innovation and creativity play a very central role in the emergence of a visible, increasing inter-dependence between technology and capital accumulation in the recent years. The most important of these innovations, in Suarez-Villa's account, is the construction of what he calls 'an intangible infrastructure', that is the digital architecture of the Internet, what Lazzarato called 'numerical networks' and the World Wide Web, the recent evolution of which into the form of social media becomes, as a matter of fact, a crucial element in this scenario. Here, we have come to a question recurrently mentioned in this work, but not yet addressed; it has been argued that the digital evolution of ICTs into social media has allowed the emergence of a few examples of alternative economic models which originate from the set of dynamics here shown so far. A very influential one is provided by Michel Bauwens (2006), who argues about the peer-to-peer mode of production as a potentially revolutionary political economic form, virtually able to reshape, in his view, the unequal neoliberal business model and, in the long term, to replace it.

A similar argument is the one carried along by Bernard Stiegler who, in his recent work on political economy, claims that the digital networks offer the instruments to establish what he calls an 'economy of contribution' (Stiegler 2010), which he sees as a bottom-up model based on the same cognitive processes of 'individuation' which I have discussed above with reference to Simondon and the multitude. According to Stiegler, the digital and numerical networks effectively provided the previously missing 'associated milieu' (Stiegler 2010) for informational commodity flows – and therefore value flows; thus, value creation increasingly takes place at a pre-individual and non-material level.

However, we have to say that the aftermath of neoliberalism – if admittedly begun, as here claimed and as Arrighi's model of accumulation cycles would suggest – is indeed going to be a long process. What we could imagine is rather a transition between the neoliberal vision to a sustainable and successful model which profitably integrates social and cultural forms of production and value creation through *affect* (Negri 1999). Affect appears to be the element around which the social production of value is centred, as it is now objectified across digital media through the 'sentiment', which stands as the new 'general equivalent' (Arvidsson 2011). The crucial role is here played by *publics*, which are *networked* (Boyd 2011) and *productive* (Arvidsson and Peitersen 2011) and which from my perspective stand as a newly collective and refreshingly active form, through the digital instrument, of multitude and audiences in the digital age, made of an intersection of people who gather for social, cultural and civic purposes (Boyd 2011) on an ethical, affective ground (Arvidsson and Peitersen 2011). It is from here that a call for an 'ethical economy' emerges around the Aristotelian definition of 'ethics' as affect, which is put at work in the contemporary business model in force in the aftermath of the 2007 crisis (Arvidsson and Peitersen 2011). Economic actors in the digital sphere are increasingly being evaluated by other members on a reputational basis which is no longer subjective, rather quantitatively assessable. Social media in fact allow through 'buttons' ('likes', 'tweets', etc.) to trace a quantitative evaluation of rankings of sentiment around brands and economic actors in general, so that their economic valorization seems to be functioning through a reputation economy (Hearn 2010, Boyd 2011, Arvidsson and Peitersen 2011, Gerlitz and Helmond 2011, Gandini 2012) in a networked information society (Benkler 2006).

Conclusions

At this point it has hopefully been shown, from the perspective here taken, how different the present landscape is from when neoliberalism originated and to what extent the role of ICTs is crucial to these dynamics. It is now time to draw some conclusions on this basis.

At a superficial reading, it may appear as though there is quite nothing new from Daniel Bell's argument on post-industrial society. Actually, he was quite right in his claim about 'the cultural contradictions of capitalism' (Bell 1976), as well as his prophecy of an incoming 'post-industrial society' (Bell 1973), not to mention 'the end of ideology' (Bell 1960). The landscape we are confronted to is actually quite similar to the one described by Bell, especially in terms of an information-based and service-oriented society. However, this is not the whole story.

From a sociological point of view, in fact, things stand quite differently. The ideological affirmation of the individual in the supposedly free, neoliberal market conceived as 'all there is' (Harvey 2005) resulted actually contaminated by a postmodern/post-Fordist rationale deeply infused with aesthetics. In other words, at the same time in which the neoliberal economic paradigm became hegemonic in society and politics, it had to face an ensemble of influences which resulted in increasingly 'immaterial', social and cultural forms of value production (Lazzarato 1996a, Negri 1999) which played with the contradictions of the neoliberal paradigm. It appears thus as though, after the plastic success of capitalism over socialism and communism, neoliberalism paradoxically got trapped into its own ideology constituted of material ends, eroded from within by immaterial instances.

This contribution has attempted to demonstrate to what extent neoliberalism, post-Fordism and postmodernity from the late 1970s up to the late 2000s, can at this point be seen as three angles of a same triangle within which ICTs in the early 1980s played the role of catalyst of neoliberalism through broadcast media and now, after a decade of the 2000s has already gone, are playing the same and inverse role against neoliberalism, through digital social media. The communication model has effectively become the actual mode of production in the 'meta-productive', self-referential process just shown, where ICTs in fact provided a technoscientific 'intangible infrastructure' (Suarez-Villa 2009) through which new economic models, potentially alternative to a neoliberal ideology can, at least ideally, develop extensively.

And still technoscience is an essential perspective through which all the processes discussed in this work can be observed. The digital networked technology seems to represent a sort of technoscientific revolution which, *mutatis mutandis*, can be compared only to Gutenberg's invention of printing or the Industrial Revolution, not only in terms of machinery but also in terms of epistemology. As Bernard Stiegler (1993) argues, human evolution is not only technical but explicitly *techno-logical*, that is connoted by an indissoluble combination of *teknè* and *episteme* (Stiegler 1993). Drawing again from Stiegler, it has not been claimed here that ICTs and technoscientific evolution triggered any anthropological mutation of the human species. In his view, it is the technical evolution that allows humans to evolve, so that technics become techno-logies (*teknè* plus *logos*) through humans and, quite explicitly, it is precisely this techno-logical potential which allows the primitive Man to become a Human, a Sapiens (Stiegler 1993). This epistemological process is analogous to the one which has been highlighted in this account, using the evolution of ICTs as our 'technics' becoming 'technologies'. This combination seems to be nowadays, perhaps as never before, by no means inextricable and the possible incoming scenarios, above all about a potentially new economic model or vision, will necessarily contemplate a discussion on media.

References

Appadurai, A. 1986. *The Social Life of Things: Commodities in Cultural Perspective.* Cambridge: Cambridge University Press.

Arrighi, G. 1994. *The Long Twentieth Century*. London: Verso.

Arvidsson, A. 2006. *Brands. Meaning and Value in Media Culture*. London: Routledge.

Arvidsson, A. 2009. The ethical economy: Towards a post-capitalist theory of value. *Capital & Class*, 33(1), 13–29.

Arvidsson, A. 2011. *General sentiment – How value and affect converge in the information economy*. [Online]. Available at: http://ssrn.com/abstract=1815031 [accessed: 28 September 2011].

Arvidsson, A. and Peitersen, N. 2011. *The Ethical Economy*. [Online. Forthcoming with Columbia University Press]. Available at: http://www.ethicaleconomy.com/info/book [accessed: 17 July 17 2011].

Baudrillard, J. 1981. *For a Critique of the Political Economy of the Sign*. St. Louis: Telos Press.

Bauwens, M. 2006. The political economy of peer production. *Post-Autistic Economics Review* [Online], 37: 33–44. Available at: http://www.paecon.net/PAEReview/issue37/Bauwens37.htm [accessed: 4 June 2011].

Bell, D. 1960. *The End of Ideology*. Glencoe, IL: Free Press.

Bell, D. 1973. *The Coming of Post-Industrial Society*. New York: Basic Books.

Bell, D. 1976. *The Cultural Contradictions of Capitalism*. New York: Basic Books.

Benkler, Y. 2006. *The Wealth of Networks.* Princeton, NJ: Princeton University Press.

Boyd, D. 2011. Social network sites as networked publics: Affordances, dynamics and implications, in *Networked Self: Identity, Community and Culture on Social Network Sites*, edited by Z. Papacharissi. New York: Routledge, 39–58.

Christopherson, S. 2008, Beyond the self-expressive creative worker: An industry perspective on entertainment media. *Theory, Culture & Society*, 25(1), 73–95.

Debord, G. 1967. *The Society of the Spectacle*. New York: Zone Books.

Drucker, P. 1942. *The Future of Industrial Man*. New York: John Day.

Florida, R. 2002. *The Rise of the Creative Class*. New York: Basic Books.

Fumagalli, A. 2007. *Bioeconomia e capitalismo cognitivo. Verso un nuovo paradigma di accumulazione*. Roma: Carocci.

Gandini, A. 2012. *Reputation economy in creative labour markets: A position paper.* Paper presented at the 28th EGOS Colloquium, Helsinki, July 5–7.

Gerlitz, C. and Helmond, D. 2011. *Hit, Link, Like and Share. Organizing the social and the fabric of the web in a Like economy*. Conference Proceedings, Digital Methods Initiative, Winter Conference, 24–25 January. Available at: http://www.annehelmond.nl/2011/04/16/paper-hit-link-like-and-share-organizing-the-social-and-the-fabric-of-the-web-in-a-like-economy [accessed: 6 December 2011].

Hardt, M. and Negri, A. 2000. *Empire*. Cambridge, MA: Harvard University Press.

Hardt, M. and Negri, A. 2004. *Multitude: War and Democracy in the Age of Empire*. London: Hamish Hamilton.
Harvey, D. 1989. *The Condition of Postmodernity*. Oxford: Blackwell.
Harvey, D. 2003. *The New Imperialism*. Oxford: Oxford University Press.
Harvey, D. 2005. *A Brief History of Neoliberalism*. Oxford: Oxford University Press.
Hearn, A. 2010. Structuring feeling: Web 2.0, online ranking and rating, and the digital 'reputation' economy. *Ephemera*, 10(3/4), 421–38.
Hesmondhalgh, D. 2002. *The Cultural Industries*. London: Sage.
Jameson, F. 1984. Postmodernism, or, the cultural logic of late capitalism. *New Left Review*, 146(3), 53–92.
Klingmann, A. 2007. *Brandscapes: Architecture in the Experience Economy*. Cambridge, MA: MIT Press.
Lazzarato, M. 1996a. *Videofilosofia. La percezione del tempo nel Postfordismo*. Roma: Manifestolibri.
Lazzarato, M. 1996b. Immaterial labor, in *Radical Thought in Italy*, edited by P. Virno and M. Hardt. Minneapolis: Minnesota University Press, 133–47.
Lazzarato, M. 1997. *Lavoro immateriale*. Verona: Ombre Corte.
Lazzarato, M. 2004. *Les révolutions du capitalisme*. Paris: Les Empêcheurs de Penser en Rond.
Lury, C. 2004. *Brands. The Logos of Global Economy.* London: Routledge.
Mandel, E. 1975. *Late Capitalism*. London: Humanities Press.
Marx, K. 1973. *Grundrisse*. London: Allen Lane.
Mezzadra, S. 2000. *Operaismo, in Enciclopedia del pensiero politico. Autori, concetti, dottrine*, edited by R. Esposito and C. Galli. Roma-Bari: Laterza, 497–8.
Michael, M. 2006. *Technoscience and Everyday Life*. Milton Keynes: Open University Press.
Mouffe, C. 2008. *Critique as Counter-Hegemonic Intervention*. [Online]. Available at: http://eipcp.net/transversal/0808/mouffe/en [accessed 16 July 2011].
Moulier Boutang, Y. 2007. *Le Capitalisme Cognitif. La Nouvelle Grande Transformation*. Paris: Editions Amsterdam.
Negri, A. 1999. Value and affect. *Boundary*, 26(2), 77–88.
Negri, A. 2008. *Reflections on Empire*. Cambridge: Polity.
Pratt, A. 2008. Creative cities. The cultural industries and the creative class. *Geografiska Annaler: Series B, Human Geography*, 90(2), 107–17.
Reich, R.B. 1992. *The Work of Nations: Preparing Ourselves for 21st Century capitalism*. New York: Knopf.
Rullani, E. 1998. Dal fordismo realizzato al postfordismo possibile: la difficile transizione, in *Il postfordismo. Idee per il capitalismo prossimo venturo*, edited by E. Rullani and L. Romano. Milano: Etas.
Simondon, G. 1989. *L'individuation pshychique et collective*. Paris: Aubier.
Stehr, N. 1994. *Knowledge Societies*. London: Sage.
Stiegler, B. 1993. *Technichs and Time*. Stanford, CA: Stanford University Press.

Stiegler, B. 2010. *For a Critique of Political Economy*. London: Polity.
Suarez-Villa, L. 2009. *Technocapitalism.* Philadelphia: Temple University Press.
Virno, P. 2004. *A Grammar of the Multitude*. New York: Semiotext(e).
Wright, S. 2002. *Storming Heaven: Class Composition and Struggle in Italian Autonomist Marxism*. London: Pluto.

PART 2
Neoliberalism, technoscience and humanity

Chapter 4

The end of history and the search for perfection. Conflicting teleologies of transhumanism and (neo)liberal democracy

Simone Arnaldi

Introduction

This chapter deals with a problematic triad. The first component of this trio is neoliberalism, to the extent it can be considered a regime in which technological innovation is defined, pursued, and achieved according to a (limited) set of rather stable and homogeneous rules and practices. The second component is transhumanism or posthumanism, as I will use the two terms as synonyms (see below, footnote 1). Transhumanism here is referred to as both a worldview, with roots dating back at least to the early twentieth century, and an upcoming movement which promotes ideas directed towards overcoming the limits to the human condition by virtue of the technological enhancement of the mind and the body (Coenen 2010). While the first two components of the trio are 'choruses', the third element of my triad is a 'soloist': Francis Fukuyama, prominent advocate of liberal democracy and economics, eminent critic of transhumanism and human enhancement.

A specific problem with triads was taught by Georg Simmel around one century ago ([1908] 1950): triads are not simply a matter of comparison between their (isolated) components and their characteristics, but also of the (uneven and multiple) relationships between their parts. Therefore, it is the relationships between my three components that this chapter tentatively examines.

In particular, a paradox is apparent: a vision of the future human society is central to the three positions I examine. These visions are not simply a speculative exercise, but they define a *telos*, a perfected state for human society to be actively pursued. This chapter argues that focusing on these 'teleological definitions' of liberal democracy (Fukuyama), market (neoliberalism), and human enhancement (transhumanism) may offer a useful but tentative perspective to examine how convergences and divergences between the components of this triad are articulated and organized. This viewpoint will be used to revisit the controversy over human enhancement that saw Francis Fukuyama among its protagonists. Indeed, it is suggested that this controversy may be reinterpreted as a conflict of teleologies, which revolves around the different notions of 'perfection' and 'perfectibility' outlined in these lines of thought.

The next section of the chapter begins this discussion by describing the convergent character of what are here defined concisely as neoliberal and transhumanist teleologies. The notions of selfhood and market will be introduced as two major joints of neoliberal and transhumanist discourses, and as two crucial dimensions of the visions of man and society they both outline. Then, the triad's third component is presented and Fukuyama's argument against transhumanism and human enhancement is illustrated. As cursorily mentioned above, I will return to this well examined topic from what I think is an unusual perspective. Indeed, the literature seems to disconnect Fukuyama's positions concerning human enhancement and the author's work on political philosophy, namely his ambitious philosophy of history centred on liberal democracy, which is developed in his works on the 'end of history' thesis. I will examine the nexus between these two bodies of work that Fukuyama himself establishes and I will highlight that, according to Fukuyama, Man is by nature a political animal and posthumanism is first and foremost a threat to the natural *telos* of human political history. Eventually, a third section will discuss the convergences and differences of these three teleologies. Their different interpretation of the notion of perfectibility will be introduced as the crucial issue around which this controversy revolves. What is the subject of perfectibility, the limits to the perfecting action, and the point of time when perfection is reached, articulate the junctures and the fractures between the three teleologies and the three corresponding components of the triad. Eventually, after recapitulating the major points of convergence and divergence, it is noticed how these teleological arguments share the two features of normativity and disclosure and what are the consequences on the dialogue between the opponents presented in this chapter.

Convergent teleologies: Neoliberalism and transhumanism

Pushing the boundaries: Neoliberalism's nested teleologies

As the characteristics of neoliberalism and their connections with technoscience are analysed thoroughly throughout the book, this section recapitulates briefly some broad features of neoliberalism, which appear particularly relevant to support what the introduction labels as its 'teleological interpretation'.

As noted elsewhere (see Introduction, Pellizzoni and Ylönen this volume) the existence of an homogenous and coherent 'neoliberal model' is questioned; accordingly, more variegated and processual notions like, for example, neoliberalization (Brenner, Peck and Theorore 2010) or neoliberal globalization (Moore et al. 2011) are instead proposed to interpret the 'contemporary processes of market-oriented regulatory restructuring' (Brenner, Peck and Theorore 2010: 182). Therefore, highlighting some broad tendencies in the neoliberal understanding of economics and politics beyond the local variations of political-economic arrangements and the institutional configurations in which notions and concepts are embodied, is surely a matter of simplification. However, some unity appears, at least at a very general level.

Broadly speaking, neoliberalism advocates, supports and implements a set of policies aimed at reorganizing the relationships between individuals, markets and States. The pillar of this reorganization is the idea that 'human well-being can best be advanced by liberating individual entrepreneurial freedoms and skills' (Harvey 2005: 2) in a free market environment. Therefore, the common core of these policies is the promotion of market-based solutions to a broad range of issues. Despite the fact that they both share the assumption of a central importance of 'free market', neoliberalism differs

> from classical political liberalism by renouncing the passive notion of a laissez-faire economy in favor of an activist approach to the spread and promotion of 'free markets'. Contrary to classical liberalism, neoliberals have consistently argued that their political program will only triumph if it becomes reconciled to the fact that the conditions for its success must be constructed. (Lave, Mirowski and Randalls 2010: 661)

The expansion of 'market relations into traditionally public arenas such as healthcare, education, and environmental management' (Lave, Mirowski and Randalls 2010: 661) is hence a matter of strategy and policy implementation, rather than in the nature of things.

The market is therefore considered as an artefact, an object of active construction, the result of 'regulatory restructuring' and 'market-based problem solving' strategies. Because market, and market relations, especially competition, are seen as the result of this coordinated policy action, neoliberalism advocates the paradoxical 'mobilization of state power in the contradictory extension and reproduction of market(-like) rule' (Birch 2006: 4).

This paradox is a paradox of naturalization that affects also the organization and practice of science.

> On the one hand practices (e.g., patenting) must encourage and promote innovation and the capacity of commercial actors to enter (i.e., make) new markets so that the market can spread in new areas of life. On the other hand innovation has to be represented as a natural process in which 'fitness' (i.e., success) is (re)presented as a consequence of inherent and endogenous competitiveness thereby justifying and naturalising a specific set of practices. (Birch 2006: 2)

By characterizing this push towards market economy in normative terms, neoliberalism describes therefore a teleological movement towards the Market.

In other words, neoliberal discourse and policy advance a complex transformation process aimed at promoting free market as an ultimate standard of economic and social organization. Such a deliberate transformation entails 'a certain concept of what man is or should be' (Bárd 2010: 75). In Imre Bárd's words:

> [b]y extending the logic of economics into almost every domain of life in an increasingly globalized and turbulent world the images of an enterprise culture and that of the entrepreneurial, flexible self were born. […] The image of the enterprising individual is premised on a view of the self as autonomous, choosing, rational; someone who pursues its own life-plans according to its own values and priorities. (Bárd 2010: 75; see also Bárd this volume)

The extension of this sort of 'capitalistic rule' to the self recasts

> a theory of *homo oeconomicus*, but the *homo oeconomicus* [in neoliberalism] is absolutely not an exchange partner. The *homo oeconomicus* is a businessperson and a businessperson of him/herself […], being for oneself one's own capital, being for oneself one's own producer, being for oneself one's own source of income. (Da Silva Medeiros 2006: 1, quoting Foucault)

Individual strategic calculation in the market is therefore directed to increasing this biological capital, to increasing return on the investment made on one's own body; (see also Pellizzoni and Ylönen this volume). Referring to the work of Nikolas Rose, Bárd notices that,

> at the very moment when countless accounts of the passing and demise of the image of the self as stable, unified and autonomous emerge in philosophy and social theory regulatory practices seek to govern individuals in a way more tied to their 'selfhood' than ever before, and the ideas of identity and its cognates have acquired an increased salience in so many of the practices in which human beings engage. (Bárd 2010: 76)

Therefore, two 'nested teleologies' apparently flourish in neoliberalism. The first is a 'systemic' one, and it concerns the goal of a society fully shaped by the market as an institution encompassing all aspects of human life. The second one is nested in the former, and it is centred on the 'individual': a teleological view of the subject, whose agency, identity, and even self-fulfilment is tied to the individual capacity of strategic calculation, planning, and design of a 'life project'. Individualization, autonomy, rationality, reflexivity, responsibility are characteristics attached altogether to these individualized subjects motivated by the desire of (and the capacity for) self-realization.

Posthumanism and the search for perfection

Transhumanism or posthumanism is both a worldview, with roots dating back at least to the early twentieth century, and a social movement which promotes ideas directed towards overcoming the limits to the human condition by virtue of the technological enhancement of the mind and the body (Coenen 2010, Russo 2006).[1] In the recent past, the rise of information technology, biotechnology, artificial intelligence and robotics, and nanotechnology, as well as the ability they donate to study and manipulate life and the material world, have popularized and increased the prominence of posthumanism and human enhancement (HE) in ethical inquiry and advice (e.g. Savulescu, Kahane and ter Meulen 2010), in S&T policy, like the US National Science Foundation (NSF) programme on Converging Technologies (e.g. Roco and Bainbridge 2001), in multidisciplinary and international scientific collaboration, like the projects ENHANCE (http://www.enhanceproject.org/) and EPOCH (http://www.epochproject.com) sponsored by the European Union and the report by Allhoff et al. (2009) sponsored by the NSF.

While examples of HE are already part of our everyday life (e.g. cosmetic surgery) and, in general, of different spheres of social life (e.g. physical enhancement in sport, the use of certain drugs as study aids or for military purposes), the discourse of transhumanism casts the technologically-enabled human capacity of self-altering against technological fields that are new, emerging or still visionary, connecting the actual present to hypothetical futures by assuming the potential of such evolving technological fields to radically alter human corporeality and intellect. No matter if these technologies are targeted to cognitive, physical and mood enhancement (Savulescu, Kahane and ter Meulen 2010), they blur the

---

1 Equating posthumanism and transhumanism is a contentious matter and this chapter does equate the two notions mostly because transhumanism is self-described (Bostrom 2005) and is described by others as a posthumanist movement (Coenen 2010). Indeed, the quest for overcoming the limits of our current human conditions enrols transhumanism among the schools of thought contesting humanistic ontologies and politics, and their attempts to essentialize human difference. The present chapter details what the features of this contestation are. However, a quote from Twine can summarize what the very specific transhumanist interpretation of posthumanism is. 'Transhumanist take things literally. Their supersession of humanism is material in a specific way. When they talk of posthumans they are imagining a human materially modified' (Twine 2010: 181). Posthumanism is much broader than this, as it spans across literary studies (Hayles 1999), animal studies (Twine 2010), feminism (Haraway 1991), philosophy (Bárd 2010), and science and technology studies (Latour 2005). Considerable efforts are being made to demarcate the boundaries between transhumanism and posthumanism (Frabetti 2004, Hayles 2008) or to enrol posthuman theorists as antecedents, sources of inspirations, and allies (Bostrom 2005, Munkittrick 2009). While the encounter of humanity and technology in authors like Latour and Haraway simultaneously creates a posthuman subject and dissolves human agency, transhumanism apparently does not escape '"West's" escalating dominations of abstract individuation, an ultimate self, untied at last from all dependency, a man in space' (Haraway 1991: 151). In Twine's terms, transhumanism is apparently 'hyper humanistic' as it 'remains faithful to a human both individualised and rational [and] it extends these both' (Twine 2010: 182) to the directional control and rationalization of the human body and the evolution of the species. Though the features of the 'transhumanist subject' is crucial to my discussion of the links between transhumanism and neoliberalism, a full exploration of what this notion implies for the controversial relationships between transhumanism and posthumanism is however far beyond the goal of this chapter.

boundaries of therapeutic interventions and actions, as they are aimed to bring about improvements extending beyond therapy (see also Bárd this volume). As an exemplary set of HE technologies, Coenen et al. (2009) list: neuroimplants that provide replacement sight or other artificial senses, drugs that boost brain power, human germline engineering and existing reproductive technologies, nutritional supplements, new brain stimulation technologies to alleviate suffering and control mood, gene doping in sports, cosmetic surgery, growth hormones for children of short stature, anti-ageing medication, and highly sophisticated prosthetic applications that may provide specialized sensory input or mechanical output.

The fundamental reason given in favour of enhancement is the belief that augmentation of physical, intellectual and emotional capacities will grant greater freedom and improve the quality of life, 'overcoming aging, cognitive shortcomings, involuntary suffering' (Humanity Plus 2011). Enhancements have been hence considered even as a 'moral obligation' and a 'duty to enhance' has been affirmed by some ethicists (e.g. Harris 2007). In transhumanist discourse, these individual enhancements are related to and framed by broader visions of an enhancement of collective performance and even of fundamental transformations of humanity, which is set to overcome the natural limitations of human species (e.g., Bostrom 2003). Examples of these 'grand visions' of enhanced societies include the indefinite postponement of death and ageing (De Grey 2007), the transition from flesh to data thanks to the digitalization of our brains and bodies (Bainbridge 2004), and even the colonization of the universe in the form of advanced robots (Kurzweil 2005).

Like neoliberalism, posthumanism defines therefore a two-fold teleology: on the one hand, enhancements are considered as a path to individual betterment beyond species-typical characteristics and constraints, resulting into new dimensions of existence and self-fulfilment; on the other hand, the aggregate effect of these individual choices promises to radically transform human societies and, in the most far-fetching visions, the place of humanity in the universe.

With no ambition to offer a comprehensive overview of posthumanism's historical antecedents, intellectual roots, and social and ethical implications, this section of the chapter points out the major ways in which transhumanism links the individual and societal dimensions of its teleology.

This connection is generally framed in terms of a liberal political stance leaving enhancement to individual choice (e.g. Agar 2004), as the 'wide personal choice over how they enable their lives' (Humanity Plus 2011) is part of the rights of individuals (e.g. Harris 2007, Canton 2006). A clear example of this attitude to privilege individual choice over societal constraints is the discussion on the issue of reproductive liberty. In the controversy over the so-called 'designer babies', an 'iconic signifier' (Franklin 2006) of a set of heterogeneous techniques for genetic screening (like preimplantation genetic diagnosis) or engineering (like germline therapy), several commentators supported a 'liberal eugenics', acknowledging to parents alone the right to make decisions about their future offspring (Agar 2004, Glover 2006). Any restrictions to be imposed by the state on individual's reproductive decisions would be legitimate only if the decisions parents make would clearly be harmful to their direct descendants.

Interestingly enough, the deep social transformations that posthumanism envisages are an emergent result of these individual decisions and actions. Despite the sweeping nature of these prospected changes, this 'individualized utopianism' powered by technology has abandoned any 'larger social project' (Bárd 2010: 32), leaving social and political futures to be created *as an aggregate result* of personal choices. Posthumanist rhetoric coherently often sees the market as an efficient and satisfactory mechanism to promote an efficient and effective social coordination of these individual actions.

This is true both in the sense of adequately supporting the development of HE technologies and of satisfactorily steering its social consequences. The market is indeed considered appropriate both for driving research and investment in a society 'accustomed to the search for personal improvement' (Michelson 2006: 51; see also Amant 2006: 198) and for governing potential unintended effects of human enhancements. On the one hand,

> [t]here is a broad scope of bodily features, physical states (or deficiencies) and behavioural particularities that until now would not have been regarded as a disease or abnormality but simply as a variety of human nature (e.g., variations in body height, a person's sociability, stress resistance, self-confidence). (Coenen et al. 2009: 59)

Medicalization is a path to marketization of these once invariable human conditions. The cases of Viagra, in vitro fertilization, and preimplantation genetic diagnosis exemplify how new medical markets are created or expanded. This tendency is particularly relevant for drugs like, for example, Ritalin and Adderall, and other stimulants that have been developed for neurological or psychological treatment, which have opened entirely new (and often illegal) markets thanks to their 'dual use' beyond the clinical problem they originally may have been developed for. Enhancement in this case happens as a consequence of a clearly non-medical use of a drug developed for a medical objective, like for students seeking to increase (illicitly) their concentration.

On the other hand, 'market-based problem solving strategies' are formulated as an answer to the governance issues raised by HE technologies. An exemplary case is the debate on issues of distributive justice potentially connected to the development and diffusion of HE technologies. This is nothing new, as

> many technological developments have led to new groups of marginalised people and to new inequalities. There is no reason in today's political realities why this would be different if the human body became the newest frontier of commodification. As much as HE technologies will become an enabling technology for the few, it will become a disabling technology for the many. (Coenen et al. 2009: 61; see also e.g. BMA 2007; Turner and Sahakian 2006)

This argument has also a variant focusing on international inequalities between rich and poor countries, whose gap is feared to be augmented by the uneven capacity to acquire cutting-edge enhancement technologies. This thesis prompts a response from the defenders of enhancement, who return to the market to find a solution to potential inequalities. Along these lines, several authors affirm that the free market itself can provide affordable access to enhancement, as increased efficiency in the production and delivery of the technology under market competition reduce prices (e.g. Stock 2002). In Harris's words, excluding market mechanisms from the equation is first and foremost 'doubtful economics and doubtful policy' (Harris 2007: 30).

Convergent futures?

Although the link between neoliberalism and transhumanism does not appear to be a popular topic, the scientific literature examined the relationships between these two schools of thought. Transhumanism is considered polemically as a vehicle through which 'some of the most questionable aspects of capitalist ideology' pave their way to the twenty-first century and it is accused of proposing an interpretive framework of the 'momentous changes in human life and culture that advanced technologies make possible' that is 'too narrow and ideologically fraught with individualism and neoliberal philosophy to be fully up to the task' (Hayles 2008). Hughes (2006) describes a more heterogeneous (bio)political landscape of the transhumanist movement, but acknowledges the importance of 'neoliberals and market anarchists' as prominent advocates for technoutopianism in the 1970s and 1980s and of the 'entrepreneurial neoliberalism' developed in the Silicon Valley as a 'natural home' for this strain of transhumanism. More consistent analysis can be found in the work of Bárd (2010) and Birch (2008). The latter comments the 'incorporation of economistic thinking' in the transhumanist framing of bioethics following the 'normative logic' that is common to the different varieties of neoliberalism, that are characterized by a totalizing view of 'market-based institutions that encourage individualistic behaviour' of rational and self-interested subjects (Birch 2008: 1). For Bárd (2010), this understanding of the self implies that transhumanism lacks a societal, collective horizon for its transformative project, which can be seen as an aggregate result of individual (and individualistic) choices.

The notions of selfhood and agency on the one hand, and of (free-)market on the other hand seem therefore two relevant points of convergence between neoliberalism and transhumanism. The preliminary notes of this chapter add to this understanding the acknowledgement that both neoliberalism and transhumanism develop a double teleology, for the self and the overall society. These directional prospects converge in describing an expanded space for human freedom respectively in terms of overriding the limitations set by the State (neoliberalism) and by Nature (transhumanism).

Accordingly, a few major similarities are apparent.

Like neoliberalism, transhumanism aspires to extend the space of human decision and to remove bonds to human choice, presuming that the liberation of human agency is the condition of self-fulfilment and well-being and that the exercise of this agency is centred on the individual, as a self-contained, autonomous, reflexive, rational entity.

Like neoliberalism, transhumanism invites individuals to capitalize on their physical and mental capacities to pursue autonomously their own path to self-realization and it presses individuals to augment these capacities to improve their efficiency and efficacy in the quest.

Like neoliberalism, transhumanism regards free market as the proper framework for achieving this liberation and for fully exploiting the potential of individuals. If the primacy of the market is definitional of neoliberalism, 'the strong relationship between libertarianism and the growing transhumanist milieu continues' (Hughes 2006: 302), though it is not uncontested.

Like neoliberalism, transhumanism adopts an active stance on the promotion of their beliefs, theories, and theses: what they see as a culmination of the human condition is not either a gift or a historical necessity, but an objective to achieve. Like neoliberalism, transhumanism is hence intrinsically political.

Therefore, it is not surprising that controversies appear when human enhancement and its consequences are discussed against the broader picture of the forms of human polity.

The world's most dangerous idea? Revisiting Fukuyama vs. transhumanism

Francis Fukuyama is a prolific writer, who is alternatively defined an economist, a sociologist, a philosopher, depending on the source. His work is a multidisciplinary venture crossing disciplinary boundaries and topics, spanning from political philosophy (Fukuyama 1989, 2006), to social change caused by the ICT revolution (Fukuyama 2000a), trust in society (Fukuyama 1995) and, of course, human enhancement.

A common opinion is that Fukuyama is to be enlisted altogether in the neoliberal camp. This opinion unites commentators, bloggers, and academic scholars (Scholte 2000, Giroux 2004) and, typically, it unites them in a critique to neoliberal ideology and its alleged supporters. However, a closer look at Fukuyama's broad yet intertwined body of work can offer a different understanding of the relationships between Fukuyama, who is certainly an advocate of liberal economics, and neoliberalism.

Although a comprehensive discussion of such a link is far beyond the goals of this chapter, it can be noticed that central to this divergence is seemingly Fukuyama's understanding of liberal democracy. In particular, though free market economics are crucial in Fukuyama's view of the liberal democratic polity, the author warns against a totalizing view of the market as a general principle for organizing social life. In his own words,

> If the institutions of democracy and capitalism are to work properly, they must coexist with certain premodern cultural habits that ensure their proper functioning. Law, contract, and economic rationality provide a necessary but not sufficient basis for both the stability and prosperity of postindustrial societies; they must as well be leavened with reciprocity, moral obligation, duty toward community, and trust, which are based in habit rather than rational calculation. (Fukuyama 1995: 11)

In other words, Fukuyama refuses the reduction of the self to a self-interested, calculating and completely autonomous subject and contests the totalizing logic of the market.

It is interesting to notice that Fukuyama's normative view of liberal democracy is equally important in defining his position vis-à-vis HE, which qualifies this author as one of the harshest critics of transhumanism, the 'world's most dangerous idea' (Fukuyama 2004).

Fukuyama's work on HE develops in the framework of an even broader interest in the governance of biotechnology, also as part of his affiliation to the US President's Council on Bioethics (PCBE). With no ambition to offer a comprehensive picture of this thought, the following sections illustrate the original line of argumentation that emerges in his writings and establishes a connection between the author's work on political philosophy, namely his ambitious philosophy of history centred on liberal democracy, which is developed in his works on the 'end of history' thesis, and the critique to transhumanism. Indeed, the literature seems to neglect this junction; this chapter examines instead the nexus between these two bodies of work that Fukuyama himself establishes.

HE and democracy: Preliminaries

The consequences of human enhancement and, more generally, of the 'biotechnology revolution' (Fukuyama 2002a) on democracy are a central concern in Fukuyama's writings, both in the works he authored and the opinions of the PCBE he collaborated to draft.

Fukuyama's concerns focus on the potential disruptive impact of the unanticipated consequences of HE on the human polity. Regarding this, Fukuyama warns that, though individual choices for biotechnical 'self-improvement' might be defended or at least not objected to on a case-by-case basis, their aggregated effects are likely to produce 'what the economists call "negative externalities"' (PCBE 2003: 283). The complexity of human mind and body on the one hand and the complexity of human societies on the other raise dramatically the potential for adverse effects of enhancing applications of genetic engineering. 'If the problem of unintended consequences is severe in the case of nonhuman ecosystems, it will be far worse in the realm of human genetics' (Fukuyama 2002b: 30). 'Modifying any one of our key characteristics inevitably entails modifying a complex, interlinked package of traits, and we will never be able to anticipate the ultimate outcome'

(Fukuyama 2004). Unrestricted enhancements 'offer a classic example of the Tragedy of the Commons, in which advantages sought by individuals are nullified, or worse, owing to the social costs of allowing them to everyone' (PCBE 2003: 66).

Regulation (Fukuyama 2002a, Fukuyama and Furger 2006) is the answer to the potential adverse consequences of HE. However, regulation cannot be 'self-regulation' of the scientific community. 'Scientists *qua* scientists have no special authority to make ethical or political judgments about the ends of their scientific research' and '[a]ll scientific research is *ultimately* subject to rules set not by scientists, but by the broader political community. [...] Indeed, in a democracy, the people are sovereign not only over science, but over every other field of activity' (Fukuyama and Furger 2006: 42).

Therefore, 'it is acceptable (and arguably necessary) for a liberal polity to place limits on individual liberty' (PCBE 2003: 283) to research and develop HE applications and to use HE applications, when their aggregated, unintended consequences may be disruptive for the existence of the polity itself.

It has been correctly said that the beacon of Fukuyama's arguments against HE is a strong notion of human nature, which gained to Fukyama the label of 'bioconservative' (Hughes 2006). As Imre Bárd puts it: 'Fukuyama's argumentation consists of three basic points: 1: there is something we can call "human essence"; 2: only those beings or individuals who possess this essence are endowed with inherent value and inviolability; 3: transhumanist aspirations would demolish or alter this essence' (Bárd 2010: 47). This 'human essence' is therefore granted by Fukuyama a normative binding status that is constitutive of political subjectivity.

Bárd follows asking what this 'human essence' or 'human nature' is about and his discussion of Fukuyama's position is valuable to summarize an apparent tension in Fukuyama's arguments against *our posthuman future*. 'Human nature is the sum of the behaviour and characteristics that are typical of the human species, arising from genetic rather than environmental factors' (Fukuyama 2002a: 130). Fukuyama acknowledges that this 'typicality' is a contingent statistical construct, based on 'a snapshot of the species at one particular moment of the evolutionary time' (Fukuyama 2002a: 152). As Bárd notices, this 'biological essentialism' seemingly ends up in a paradox: 'Fukuyama effectively continues the very line of thought from which he intended to save us all' (Bárd 2010: 48), reducing human nature to a biological, material, pliable entity, like transhumanism does. From this perspective, it is impossible to argue in favour of a normative human nature and to answer the question 'why not seize this power [of enhancing ourselves], indeed?' (Fukuyama 2002a: 153). The normative value of this descriptive typicality is brilliantly challenged by John Harris, who recalls Richard Dawkins and warns of

> the dangers of making a fetish of a particular evolutionary stage. If our ape ancestor had thought about it, she might have taken the view adopted by so many of our contemporary gurus like Leon Kass, Michael Sandel, George Annas, Francis Fukuyama, and many others, that there is something special about themselves and that their particular sort of being is not only worth preserving

> in perpetuity, but that there is a duty not only to ensure that preservation, but to make sure that neither natural selection nor deliberate choice permit the development of any sort of being. (Harris 2007: 16)

To rescue the normativity of the notion of human essence, Fukuyama argues for the existence of a 'Factor X', 'Factor X cannot be reduced to the possession of moral choice, or reason, or language, or sociability, or sentience, or emotions, or consciousness or any other quality that has been put forth as a ground for human dignity. It is all of these qualities coming together in a human whole that make up Factor X' (Fukuyama 2002a: 171). 'This is what remains when individuals are stripped of all accidental properties and this Factor X is the basis of human dignity' (Bárd 2010: 49). This is certainly a qualitative leap from a statistical, descriptive typicality to a normative typicality.

While this position is criticized as 'mysticism' in disguise (e.g. Bárd 2010), less attention has been paid to understand the characteristics of Fukuyama's idea of human essence by referring to other works of this author. In my view, his discussion of transhumanism and genetic manipulation cannot be separated by his broader view of human polity and, more specifically, liberal democracy. Indeed, Fukuyama is concerned with human beings as Aristotelian 'political animals' and it is exactly the 'human political nature' that abridges descriptive and normative typicality.

Fukuyama's political philosophy is the underlying framework of his discussion about biotechnology consequences (Fukuyama 2002a: 13–15) and he establishes a direct connection between his famous view of the end of history and transhumanism, which interestingly enough closes the second edition of his famous *The End of History and the Last Man* (Fukuyama 2006: 353–4). Though Fukuyama's commentators tend to disregard this second body of writings in their interpretation and critique of Fukuyama's positions on HE, I argue that examining the theory of the 'End of history' illuminates a 'third layer' of Fukuyama's view of human nature. It is this third layer that puts human essence in the perspective of a teleological evolution of humankind.

The end of history or the end of the world (as we know it)?

Though it has been subject to some adjustments, the master narrative of the end of history, and the place that liberal democracy has in it, remains central to Fukuyama's political thought. The original 1989 article on 'The end of history' remarked that '[t]he triumph of the West, of the Western idea, is evident first of all in the total exhaustion of viable systematic alternatives to Western liberalism' following the end of the cold war and the fall of the Berlin Wall. According to Fukuyama,

> [w]hat we may be witnessing is not just the end of the Cold War, or the passing of a particular period of postwar history, but the end of history as such: that

> is, the end point of mankind's ideological evolution and the universalization of Western liberal democracy as the final form of human government. [and] the total exhaustion of viable systemic alternatives to Western liberalism. (Fukuyama 1989)

> This had concluded history, not in the trivially untrue sense of bringing the empirical events to an abrupt halt, but in as much as it had realised its goal: freedom. What we are witnessing it is not a cessation, but a culmination. (Elliott 2008: 38)

While somebody has noticed that the faith in the triumph of the US version of liberal democracy and of capitalist economy has become less adamant over time (Elliott 2008: 49–51), a position that Fukuyama considers a 'widespread misapprehension' of his thought (Fukuyama 2006: 346–7), the fundamental tenets of Fukuyama's 'weak determinism' are still in place (Fukuyama 2006: 354).

Fukuyama establishes a distinction between 'history', the 'occurrence of events' and 'History', 'a single coherent evolutionary process' (Fukuyama 2006: xiii), which is directed by both the realm of things and the realm of ideas (of consciousness). Relying largely on Hegel,[2] Fukuyama criticizes the Marxist view of historical evolution as a result of the dialectic interplay of material forces, to affirm that 'understanding the underlying processes of history requires understanding developments in the realm of consciousness or ideas, since consciousness will ultimately remake the material world in its own image' (Fukuyama 1989). The result is a view of human socio-political evolution as the product of two mechanisms of transformation, 'ideological' and 'material' (Elliott 2008: 42). On the one hand, the logic of 'modern natural sciences' drives to the capacity to exploit and transform natural resources to satisfy material desires. This is due to the cumulative logic of natural sciences, which makes historical evolution directional (Fukuyama 2006: xiii). On the other hand, Fukuyama relies on Hegel to affirm the 'struggle for recognition' as the inner motor of human action:

> human beings like animals have natural needs and desires for objects outside themselves such as food, drink, shelter, and above all preservation of their own bodies. Man differs fundamentally from the animals, however, because in addition he desires the desire of other men, that is, he wants to be 'recognized'. (Fukuyama 2006: xvi)

Borrowing the notion of *thymos* ('spiritedness') from Plato's theory of the soul, Fukuyama turns to human need for recognition and human thymotic feeling as a second driver of history.

---

2 Fukuyama's reading of Hegel is mediated by the work of the French philosopher Alexandre Kojève (1902–1968). Fukuyama explicitly declares that 'we are not interested in Hegel per se, but in Hegel-as-interpreted-by-Kojève' (Fukuyama 2006: 144).

> It is like an innate human sense of justice. People believe that they have a certain worth, and when other people treat them as though they are worth less than that, they experience the emotion of anger. Conversely, when people fail to live up to their own sense of worth, they feel shame, and when they are evaluated correctly in proportion to their worth, they feel pride. (Fukuyama 2006: xvii)

These dynamics of recognition resulted in conflicts and bloodsheds, domination and slavery, throughout human history:

> the desire to be recognized as a human being with dignity drove man at the beginning of history into a bloody battle to the death for prestige. The result of this battle was a division of human society into a class of masters, who were willing to risk their lives, and a class of slaves, who gave in to their natural fear of death. (Fukuyama 2006: xvii)

However, the advent of liberal democracy changed human history as it constitutes a discontinuity in the flow of historical events and, simultaneously, it represents the end and the culmination of 'a coherent and directional Universal History of Mankind' (Fukuyama 2006: xxiii).

This happens because the two drivers of human history converge and are mutually perfected in the liberal democratic polity. Firstly, thymotic pride is no longer directed to battle and domination, but it is 'sublimated' into mutual recognition.

> Democratic revolutions abolished the distinction between master and slave by making the former slaves their own masters. [...] The inherently unequal recognition of masters and slaves is replaced by universal and reciprocal recognition', while '*megalothymia* [the desire to be recognised superior to others] is sublimated into entrepreneurship, politics, or even leisure sports. (Fukuyama 2006: xvii–xviii, 315–16)

The granting of individual rights is fundamental in this process. Individuals have a 'thymotic pride in their own self-worth, and this leads them to demand democratic governments that treat them like adults rather than children, recognizing their autonomy as free individuals' (Fukuyama 2006: xix). Secondly, free market economics boost the transformation of the technological power provided by modern natural sciences into the capacity for unlimited material accumulation able to satisfy the human desire for goods. A 'truly universal consumer culture' that will become 'both a symbol and an underpinning of the universal homogenous state' is the arrival of the political and economic evolution of mankind. 'The end of history will be a very sad time. The struggle for recognition, the willingness to risk one's life for a purely abstract goal, the worldwide ideological struggle that called forth daring, courage, imagination, and idealism, will be replaced by economic calculation, the endless solving of technical problems, environmental concerns, and the satisfaction of sophisticated consumer demands' (Fukuyama 1989).

The success of liberal democracy is hence a consequence of the convergence of the two motors of human history, which are, in turn, fuelled by the inner motivating forces of human action: the desire for recognition and the desire for material goods. 'Liberal democracy has emerged as the only viable and legitimate political system for modern societies because it avoids either extreme, shaping politics according to historically created norms of justice while not interfering excessively with natural patterns of behavior' (Fukuyama 2002a: 14). No systemic alternative to liberal democracy is possible not only because history recorded the factual failure of other regimes, but also because this form of government is intimately consistent with human nature and, thus, it can drive human History to its just and natural end.

Posthumanism and the challenge to liberal democracy

While stating a resounding 'there is no alternative' (or TINA as the well-known acronym goes) to the question about the existence of any possible systemic alternative to liberal democracy, Fukuyama warns that his view of history is only 'weakly deterministic' and, hence, surprises are still possible. 'My historicist view of human development has always been only weakly deterministic. […] Weak determinism means that in the face of broad historical trends, statesmanship, politics, leadership and individual choice remain absolutely critical to the actual course of historical development' (Fukuyama 2006: 354).

If external enemies cannot defeat liberal democracy, there is no guarantee that the 'enemy within' may ever subvert it, thus derailing human History from the natural path to its *telos*. In particular, Fukuyama mentions the 'ability to manipulate ourselves biologically, whether through control over the genome or through psychotropic drugs, or through a future cognitive neuroscience, or through some form of life extension' (Fukuyama 2006: 353–4). According to Fukuyama, 'this particular technological future' poses a threat that is 'much more subtle than the one posed by nuclear weapons or climate change', as the potentially dangerous 'consequences of technological advance are tied up with things like freedom from disease or longevity that people universally want, and will therefore be much more difficult to prevent'.

According to Fukuyama, modern natural sciences have been working to favour the advent of liberal democracy creating the technological engine of free market economics throughout human history. However, if science and technology are directed not to enable our domination of the external natural environment, but to the modification of our human nature, they can alter the fundamental motors of human History, thus bringing to an end its directional development. Faithful to his understanding that the roots of human behaviour lie in the struggle for recognition (Fukuyama 1989, 2006), he discusses the possible consequences of altering biologically ourselves.

> Hegel believed that the struggle for recognition was a purely human phenomenon – indeed, that it was in some sense central to what it meant to be a human being. But in this he was wrong: there is a biological substrate for the human desire for recognition […] that is related to levels of serotonin in the brain. (Fukuyama 2002a: 45)

The possibility to alter the fragile balance of recognition is therefore the issue Fukuyama is most concerned of. However, while commentators usually focus on Fukuyama's warning against the danger of creating a new dominating class of 'masters' seeking to enhance themselves thus creating a new form of 'Lordship' and undermining the mutual recognition that supports the liberal democratic polity, the author's concerns regarding human biological manipulation are double-edged. Indeed, '[l]iberal democracy could, in the long run, be subverted internally either by an excess of *megalothymia*, or by an excess of *isothymia* – that is the fanatical desire for equal recognition' (Fukuyama 2006: 314). Therefore, in addition to a possible megalothymotic class of enhanced masters, Fukuyama fears 'a civilization that indulges in unbridled *isothymia*, that fanatically seeks to eliminate every manifestation of unequal recognition' (Fukuyama 2006: 314), where 'tomorrow's isothymotic passions' fuelled by the power of new genetic technologies try to overcome the limits imposed by nature and to outlaw any differences among the members of the political community. This is an opposite, and apparently contradictory, concern. However, according to Fukuyama, this search for unlimited equality challenges the functioning of liberal democracy because 'every contemporary liberal democracy does in fact differentiate rights based on the degree to which individuals or categories of individuals share in certain species-typical characteristics' (Fukuyama 2002a: 175). In other words, mutual recognition does not imply equal recognition, as all humans are not created equal in terms of their physical and intellectual capabilities and the acknowledgement of these natural differences is needed for the very functioning of a liberal democratic polity. However, biotechnology offers the tools to level (natural) differences by changing what were previously considered the given (natural) conditions of existence, thus leading to a paradox in liberal democratic policies. 'Democracy's teaching about human equality has a basis in nature but also violates human nature by pretending that unequal men are in fact equal' (Fukuyama 2000b). If

> [t]he possibility that biotechnology will permit the emergence of new genetic classes has been frequently noted and condemned by those who have speculated about the future, […] the opposite possibility also seems to be entirely plausible – that there will be an impetus toward a much more genetically egalitarian society. […] This is the only scenario in which it is plausible that we will see a liberal democracy of the future get back into the business of state-sponsored eugenics. (Fukuyama 2002a: 159)

Paradoxically, democratization of enhancement technologies does not mean therefore a healthy future for liberal democratic polities: '[t]he state and market could ensure there is full democratization [of enhancement technologies], [...] but this would only universalize the other problem of distorting human nature' (Best and Kellner 2001).

There is no alternative: Dissecting teleologies

There is no alternative, or TINA, is Margaret Thatcher's famous slogan excluding viable replacements of neoliberal economic policies. Such a statement implicitly resounds three times in the previous sections of this chapter. Firstly, neoliberalism excludes alternatives to the market as a mechanism for integrating economy and society – the latter is in turn dissolved into the individuals that are its components. Secondly, transhumanism excludes alternatives to a human future of technologically-enabled freedom. Thirdly, Fukuyama excludes alternatives to liberal democracy as the fulfilment of the existential aspiration of human beings.

These bold statements of exclusivity are based on three different teleologies that seem to have, however, two features in common. Firstly, they maintain the perfectibility of human life, thus promising an ultimate state of individual and social evolution, an arrival of a human development directed to 'perfection'. This last, perfected state does not imply a freeze of historical contingency, an 'abrupt halt' to empirical events and transformations; it is not a steady state, but rather a space for the flourishing and progress of human freedom: freedom from want and slavery (Fukuyama), from the State (neoliberalism), and from Nature (transhumanism); it is a 'culmination' or 'fulfilment' of human aspirations and not the end of the world as we know it. Secondly, the three positions claim that our future is, by and large, disclosed to us, either because we have mechanisms (the market) to predict transformation or techniques (technology) to direct by means of purposeful action historical (and even) human evolution. Similarly, the stability of human nature (Fukuyama) enables us to understand the directions and destinations of human evolution (and its unintended detours).

However, despite these parallelisms, the differences of these three teleologies become apparent when we turn to their common, but ambiguous, notion of perfectibility. This ambiguity emerges from a comparison of three major dimensions of this concept as it is developed in the different bodies of thought that have been presented: (1) what can (and should) be perfected, (2) the limits to this 'search for perfection', and (3) the 'point of time' when perfection can be reached (see Table 4.1).

Neoliberalism and Fukuyama's version of bioconservatism (Hughes 2006) share the object of perfectibility: polity. Either as a result of historical evolution or as a product of market-oriented regulatory restructuring, it is the human form of government that is subject to a progressive journey to perfection. In transhumanism, what should be perfected is nature itself: changing society is merely a consequence, an adjustment of our increased technological capacity to modify/enhance human bodies and minds. Nature and polity reverse their places when limits to perfectibility are discussed. Indeed, neoliberalism and transhumanism consider the limits to our perfecting action only as the result of a contingent, historical consensus on what should or should not be altered and the corresponding political and institutional arrangements enforcing these beliefs and representations. The imperative is therefore liberating humankind from the bonds set either by temporary social arrangements (neoliberalism) or by the transient self-understanding of an evolving 'homo sapiens' (transhumanism). On the contrary, Fukuyama considers human nature itself as the source of the limits to our perfecting action: a stable human nature is the normative set against which any policy should be measured and assessed. Also, it is human nature that founds human polity: when political regimes did not conform to human nature, either they failed to survive altogether or they dehumanized their fellow citizens.

Table 4.1 A comparison of the notion of perfectibility

	Fukuyama's bioconservatism	**Neoliberalism**	**Transhumanism**
Subject of perfectibility	Polity	Polity	Nature
Limits to perfectibility	Nature	Polity	Polity
Point of time for perfectibility	Past (human and political evolution)	Future (prospected regulatory restructuring)	Future (prospected technical capacity)

These different notions and roles of Nature and Polity lead to incompatible perspectives on the point of time for perfectibility. Neoliberalism and transhumanism cast the reach of perfection in the future: it is actively sought through, respectively, 'regulatory restructuring' or augmented technical capacity. Fukuyama thinks that perfection has already shown its face (liberal democracy), which is to be tuned, nurtured, and kept in good conditions, but whose characteristics have been already disclosed by our recent history.

Manifest foes and false friends: Some closing remarks

Detecting friends and foes

This chapter attempted to explore the relationships between some manifest enemies and to unveil some false friends. Crucial to this examination has been the focus on the teleologies underlying Fukuyama's theory of liberal democracy, the neoliberal understanding of the market, and the transhumanist interpretation of enhanced humans and societies. Also, it has been noted that these teleological definitions are not irrelevant to motivate both the convergences and the conflicts arising among these three positions.

The apparently converging *teloi* of transhumanism and neoliberalism relate to the idea of an unbounded subject, whose enhancement may be considered a sort of extreme capitalization of one's own physical and mental capacities by en entrepreneurial subject, and an individualistic vision of society. The latter is regarded as the result of individual striving for self-fulfilment, which market mechanisms are better positioned to coordinate than the State or political authority in general. Also, both transhumanism and neoliberalism adopt an active stance on the realization of their respective *teloi*: posthumanity and market restructuring are to be tirelessly constructed and they are not 'in the nature of things'. In Fukuyama's thought, human beings are first and foremost 'political animals'; their self-fulfilment depends on the belonging to a broader human polity, and especially to a specific one: liberal democracy. This is the teleological culmination of a directional Universal Human History at the intersecting of the two major drivers of human evolution: the desire for material goods and for recognition. If this directional evolution grants liberal democracy to be safe from systemic (external) alternatives, our duty is to protect the liberal democratic polity from the enemies inside. We have therefore to avoid the wreck of these untroubled prospects for humankind by limiting activities that threat the natural development of these historical drivers and that may alter what is only a 'weak determinism'. Transhumanism is one of these threats, as its promise of freedom from nature strikes the very core of the dynamics of recognition in liberal democracy, creating the possibility of either a megalothymotic class of ruling masters or a isothymotic population of servants.

This chapter has framed this vigorous clash in terms of divergent interpretations of what 'perfectibility', that is our capacity for improvement, means. This analysis of teleologies was meant to enrich our understanding of the relationships between these old, manifest foes, that is Francis Fukuyama and a broad intellectual movement concisely labelled as 'transhumanism', as well as of the convergence between the latter and neoliberalism.

In this chapter we have assisted to a curious reversal of nature and polity: what transhumanism considers as an object to be perfected (nature) defines the limits to our perfecting actions in Fukuyama's position; what can be perfected for Fukuyama (polity) is a contingent obstacle to overcome for transhumanists seeking the construction of a posthuman society. Furthermore, according to Fukuyama, history has come to an end so far, as we have already found the best form of organized political society. This means that perfection was realized with the advent of the liberal democratic model of government; it is now a matter of propagating that model and maintaining it where it is already in place. Transhumanists think instead that the time of perfection is in the future and that we will reach it through the use of new and emerging technologies that increase our transformative power and, as a consequence, the plasticity of nature.

Also, we have described how neoliberalism shares the liberating 'mission' of transhumanism by other means. Market-oriented regulatory restructuring is seen as a purposeful perfecting action targeting historically contingent socio-political arrangements that are to be changed. However, the transformative impulse of neoliberalism stops before nature. It is not the latter that is to be perfected; instead, the societal transformation envisaged by neoliberalism relies upon a specific nature of individuals, which are assumed to be autonomous, choosing, rational, pursuing their own life-plans according to their own values and priorities in a market environment.

Examining the links between neoliberalism and transhumanism was useful to unveil the relationships between what appears to be two false friends, Fukuyama's idea of liberal democracy and neoliberalism. A thorough discussion of this seemingly false friendship is far beyond the goals of this chapter and would require the introduction of a connected work, namely Fukuyama's research on the creation and role of trust in society (Fukuyama 1995) and, to a lesser extent, his study on information technologies and social order (Fukuyama 2000a). A first difference is a specific attention to the problem of inequality as a threat undermining liberal democracies in the long run (e.g. Fukuyama 2011a, 2011b). Though surely not a social democrat, Fukuyama is concerned that the careful balance of recognition and accumulation in liberal democratic societies may be disrupted by an excessive degree of internal inequality. However, the major point is surely the acknowledgement that the market alone cannot produce social resources that are essential to the long term stability of a society, like trust and social capital. Other types of associations are to be included in the picture, like hierarchical relations and communitarian ties.

Ending history or closing down the dialogue?

The analysis developed in the chapter illustrated a paradox. What fuels divergences and contrast is first and foremost the similarities between the different positions. The first one is normativity: as far as teleological arguments outline a perfect state, no matter how ill-defined, they exclude, by definition, the possible existence of equally good or viable alternatives. The second one is disclosure: along with the definition of a general goal for human development, it is assumed that we already know the consequences of our action (or inaction), including our technological choices.

They all are 'one-way perfections' and mutual exclusion derives at the very outset from the teleological definitions of liberal democracy, enhancement and market they elaborate. They do not only bring history to an end, by they also bring the possible discussion to a close before really opening it. However, a liberal posthumanity, if any is in sight, requires a posthuman liberalism, whose elaboration needs a meaningful dialogue. Like in reading a novel, starting from the end may not be the best choice.

Acknowledgement

I wish to thank Christopher Coenen, Francesca Marin, Luigi Pellizzoni and Marja Ylönen for their insightful and encouraging remarks about an earlier draft of this chapter.

References

Agar, N. 2004. *Liberal Eugenics. In Defense of Human Enhancement*. Oxford: Blackwell.

Allhoff, F., Lin, P., Moor, J. and Weckert, J. 2009. *Ethics of Human Enhancement: 25 Questions & Answers*. [Online]. Available at: http://www.humanenhance.com/ [accessed: 22 August 2011].

Amant, R. St. 2006. Information technology and cognitive systems, in *Managing Nano-bio-info-cogno Innovation. Converging Technologies in Society*, edited by W.S. Bainbridge and M.C. Roco, Dordrecht: Springer, 193–201.

Bainbridge, W.S. 2004. Progress toward Cyberimmortality, in *The Scientific Conquest of Death*. Buenos Aires: Libros en Red, 107–22. [Online]. Available at http://www.imminst.org/book/ [accessed: 9 March 2009].

Bárd, I. 2010. *Troubles with Posthumanism*. Vienna: University of Vienna.

Best, S. and, Kellner, D. 2001. *Biotechnology, Democracy, and the Politics of Cloning*. [Online]. Available at http://www.drstevebest.org/Essays/BiotechnologyDemocracy.htm [accessed: 22 August 2011].

Birch, K. 2006. The neoliberal underpinnings of the bioeconomy:Tthe ideological discourses and practices of economic competitiveness. *Genomics, Society and Policy*, 2(3), 1–15.

Birch, K. 2008. Neoliberalising bioethics: Bias, enhancement and economistic ethics. *Genomics, Society and Policy*, 4(2), 1–10.

BMA (British Medical Association) 2007. *Boosting Your Brainpower: Ethical Aspects of Cognitive Enhancements. A Discussion Paper From the British Medical Association*. [Online]. Available at http://www.bma.org.uk/ethics/health_technology/CognitiveEnhancement2007.jsp [accessed: 22 August 2011].

Bostrom, N. 2003. Human genetic enhancements: A transhumanist perspective. *Journal of Value Inquiry*, 37(4), 493–506.

Bostrom, N. 2005. A History of transhumanist thought. *Journal of Evolution and Technology* [Online], 14(1). Available at: http://jetpress.org/volume14/bostrom.pdf [accessed: 8 December 2011].

Brenner, N., Peck, J. and Theorore, N. 2010. Variegated neoliberalization: Geographies, modalities, pathways. *Global Networks*, 10(2), 182–222.

Canton, J. 2006. NBIC convergent technologies and the innovation economy: Challenges and opportunities for the 21st century, in *Managing Nano-bio-info-cogno Innovation. Converging Technologies in Society*, edited by W.S. Bainbridge and M.C. Roco. Dordrecht: Springer, 255–78.

Coenen, C., Schuijff, M., Smits, M., Klaasen, P., Hennen, L., Rader, M. and Wolbring, G. 2009. *Human Enhancement*. Brussels: European Parliament, Science and Technology Options Assessment (STOA). Available at: http://www.itas.fzk.de/deu/lit/2009/coua09a.pdf [accessed: 23 December 2011].

Coenen, C. 2010. Immagini di società potenziate dalla nanotecnologia. L'ascesa dell'ideologia postumanista del progresso estremo, in *Innovazioni in corso. Il dibattito sulle nanotecnologie fra diritto, etica e società*, edited by S. Arnaldi and A. Lorenzet, Bologna: Il Mulino, 225–58.

Da Silva Medeiros, F.N. 2006. Convergent discourses: Neoliberalism, technoscience and journalism. *Journal of Science Communication*, 5(1), 1–3.

De Grey, A. 2007. *Ending Aging: The Rejuvenation Breakthroughs that Could Reverse Human Aging in Our Lifetime*. New York: St. Martin's Press.

Elliott, G. 2008. *Ends in Sight*. London: Pluto Press.

Franklin, S. 2006. Better by design?, in *Better Humans? The Politics of Human Enhancement and Life Extension*, edited by P. Miller and J. Wilsdon. London: Demos, 86–94.

Frabetti, F. 2004. Postumano, in *Dizionario di studi culturali*, edited by R. Coglitore and F. Mazzara. Roma: Meltemi.

Fukuyama, F. 1989. The end of history? *The National Interest*, Summer [Online]. Available at: http http://www.wesjones.com/eoh.htm [Accessed 22 August 2011].

Fukuyama, F. 1995. *Trust: The Social Virtues and the Creation of Prosperity*. New York: Free Press.

Fukuyama, F. 2000a. *The Great Disruption: Human Nature and the Reconstitution of Social Order*. New York: Free Press.

Fukuyama, F. 2000b. The march of equality. *Journal of Democracy*, 11(1), 11–17.

Fukuyama, F. 2002a. *Our Posthuman Future. Consequences of the Biotechnology Revolution*. New York: Farrar, Straus and Giroux.

Fukuyama, F. 2002b. In defense of nature, human and non-human. *World Watch Magazine*, July–August, 30–32.

Fukuyama, F. 2004. Transhumanism. *Foreign Policy* (September–October). [Online]. Available at: http://www.foreignpolicy.com/articles/2004/09/01/transhumanism [Accessed 22 August 2011].

Fukuyama, F. 2006. *The End of History and the Last Man*. New York: Free Press.

Fukuyama, F. and Furger, J. 2006. *Beyond Bioethics: A Proposal for Modernizing the Regulation of Human Biotechnologies*. Washington, DC: SAIS.

Fukuyama, F. 2011a. Dealing with inequality. *Journal of Democracy*, 22(3), 79–89.

Fukuyama, F. 2011b. Left out. *The American Interest Online*, January–February. [Online]. Available at: http://www.the-american-interest.com/article.cfm?piece=906 [Accessed 22 August 2011].

Giroux, H. 2004. Public pedagogy, and the politics of neoliberalism: Making the political more pedagogical. *Policy Futures in Education*, 2(3–4), 494–651.

Glover, J. 2006. *Choosing Children – The Ethical Dilemmas of Genetic Intervention*. Oxford: Oxford University Press.

Haraway, D. 1991. A cyborg manifesto: science, technology, and socialist-feminism in the late twentieth century, in *Simians, Cyborgs and Women: The Reinvention of Nature*. New York: Routledge, 149–81.

Harris, J. 2007. *Enhancing Evolution: The Ethical Case for Making Better People*. Princeton, NJ: Princeton University Press.

Harvey, D. 2005. *A Brief History of Neoliberalism*. Oxford: Oxford University Press.

Hayles, K. 1999. *How We Became Posthuman: Virtual Bodies in Cybernetics, Literature, and Informatics*. Chicago, IL: University of Chicago Press.

Hayles, K. 2008. *Wrestling with Transhumanism*. [Online]. Available at: http://www.metanexus.net/-magazine/tabid/68/id/10543/Default.aspx [accessed: 8 December 2011].

Hughes, J.J. 2006. Human enhancement and the emergent technopolitics of the 21st century, in *Managing Nano-bio-info-cogno Innovation. Converging Technologies in Society*, edited by W.S. Bainbridge and M.C. Roco. Dordrecht: Springer, 285–308.

Humanity Plus. 2011. *Transhumanist Declaration*. [Online]. Available at: http://humanityplus.org/-learn/transhumanist-declaration/ [accessed: 22 August 2011].

Kurzweil, R. 2005. *The Singularity is Near*. New York: Viking Penguin.

Latour, B. 2005. *Reassembling the Social: An Introduction to Actor-Network Theory*. Oxford: Oxford University Press.

Lave, R., Mirowski, P. and Randalls, S. 2010. Introduction: STS and neoliberal science. *Social Studies of Science*, 40(5), 659–75.

Michelson, E. 2006. Measuring the merger: examining the onset of converging technologies, in *Managing Nano-bio-info-cogno Innovation. Converging Technologies in Society*, edited by W.S. Bainbridge and M.C. Roco. Dordrecht: Springer, 47–69.

Moore, K., Kleinman, D., Hess, D. and Frickel, S. 2011. Science and neoliberal globalization: A political sociological approach. *Theory and Society*, 40(5), 505–32.

Munkittrick, K. 2009. On the importance of being a cyborg feminist. *Hplusmagazine* [Online]. Available at: http://hplusmagazine.com/articles/politics/importance-being-cyborg-feminist [accessed: 8 December 2011].

PCBE (President's Council on Bioethics). 2003. *Beyond Therapy. Biotechnology and the Pursuit of Happiness*. Washington, DC: The President's Council on Bioethics.

Roco, M. and Bainbridge, W. 2001 (eds.) *Converging Technologies for Improving Human Performance*. Arlington, VI: National Science Foundation. Available at: http://www.wtec.org/ConvergingTechnologies/1/NBIC_report.pdf [accessed: 20 December 2011].

Russo, V. 2006. *Più che umani: la bioetica filosofica e le tecnologie del potenziamento psicofisico* [Online]. Available at: http://www.estropico.com/id308.htm [accessed: 22 August 2011].

Savulescu, J., Kahane, G. and ter Meulen, R. 2010. *Enhancing Human Capacities*. Oxford: Wiley-Blackwell.

Scholte, J.A. 2000. *Globalization: A Critical Introduction*. Basingstoke: Palgrave.

Simmel, Georg, [1908] 1950. *The Sociology of Georg Simmel*, compiled and translated by Kurt Wolff. Glencoe, IL: Free Press.

Stock, G. 2002. *Redesigning Humans: Choosing our Children's Genes*. London: Profile Books.

Turner, D. and Sahakian, B. 2006. The cognition-enhanced classroom, in *Better Humans? The Politics of Human Enhancement and Life Extension*, edited by P. Miller and J. Wilsdon, London: Demos, 79-85.

Twine, R. 2010. Genomic natures read through posthumanisms, in *Nature after the Genome*, edited by S. Parry and J. Dupré. Oxford: Blackwell, 175–95.

Chapter 5
Pre-empting the threat of human deficiencies

Imre Bárd

Introduction

The 2009 movie *Surrogates*, starring Bruce Willis in the lead role, depicts a near-future world in which everyday citizens use remotely operated humanoid robots as their surrogates. By synchronizing their nervous system with almost indestructible, forever young and able-bodied alter egos, people can avoid all the hazards, vicissitudes and accidents of the outside world. As we learn from the movie, the introduction of this kind of surrogacy brought about a much safer world, in which homicide and fatal accidents had been virtually eradicated. The freedom to experiment and to slip into invincible bodies of all shapes and sizes imaginable has nevertheless resulted in humans becoming much like captives, trapped within the comfort of their own homes, utterly devoid of technologically unmediated interaction and genuine interpersonal relationships. It is a world of constant surveillance, dominated by fear of the uncertain, which is conquered yet simultaneously reinforced with recourse to a radical technological solution, that is available on a free-market basis.

A number of science fiction works have dealt with the tension between security and social stability on the one hand and human freedom on the other. It seems as though security and stable social structures necessarily involved sacrificing individual liberties, intimacy or some other aspect of life we usually deem valuable. Aldous Huxley's *Brave New World* (Huxley 1998) is probably the paradigmatic example of a dystopian scenario built upon this antagonism. Even though *Surrogates* is by no means on par with Huxley's work in terms of literary excellence, it does point to the problem I wish to tackle in this chapter. In fact, there is an arc spreading from the *Brave New World* dystopia of a totalitarian state, which produces its citizens on a grand scale by genetic selection and brain washing, to the vision of commercially available robotic surrogates, where people are apparently free to choose.

It is the dialectic involved in the drive for security and control, which ultimately results in voluntary self-restriction, that I want to discuss, with a focus on human enhancement technologies. I argue that we may currently be witnessing the burgeoning transformation of the way enhancement measures are conceptualized and observe their transformation into preventive and pre-emptive interventions to ensure safety in a world suffused with the dread of risks and uncertainty.

What is human enhancement?

Mankind has always used culture-specific methods and practices to increase the capacities and extend the limits of human biology. These methods can range from the use of substances like caffeine or gingko biloba, through the creation and application of increasingly complex tools, up to systematic ways of disciplining and educating the mind and the body. In a certain sense the entire process of cultural evolution can be understood as a series of attempts to overcome *natural* human limitations.

However, according to some, the quest to better the human condition has reached a crucial turning point. The convergence of the fields of nanotechnology, biotechnology, information technology and cognitive science (NBIC) seems to enable never before seen degrees of intervention into matter, into life processes and into human nature (Roco and Bainbridge 2002, PCB 2003). Recent advances have effectively blurred the lines between therapeutic and enhancing interventions on the human body so that is becoming increasingly possible to use biotechnological means not just to cure disease or ameliorate suffering but also to improve upon normal, healthy functioning.

It is very difficult to clearly define what human enhancement refers to precisely, because our current form of life, the kinds of people we are and the ways of life we inhabit have been deeply and thoroughly shaped and transformed by medical practices (see also Arnaldi this volume). Interventions such as vaccination, public hygiene, the sewage system and many more have exerted a lasting influence on human health and longevity and form an important part of the history of trying to improve the human condition (Rose 2007b). Even though sanitation has played an enormous role in improving health and quality of life, it would most probably not fall under the category of human enhancement in the present discussion. Furthermore, whatever counts as an enhancement is always conditional upon what is considered to be normal in the present.

For the purposes of this chapter I am going to use Eric Juengst's definition, which considers enhancements to be interventions on the body 'to improve performance, appearance, or capability besides what is necessary to achieve, sustain or restore health' (Juengst 1998: 29). Certainly, this definition is not flawless, because health and normality are changing notions themselves. Nevertheless, it will suffice as an operational definition based on which certain practices can clearly be described as enhancements today. The off-label use of the psychopharmacological agent Modafinil to increase the wakefulness and attention-span of students, surgeons or soldiers without apparent negative side effects is one such example (Repantis et al. 2010). This substance was initially developed for the treatment of narcolepsy and related disorders but has been proven to provide an edge to healthy individuals as well. In fact, about 90% of Modafinil usage is off-label (O'Connor 2004). In the future we might see cognition enhancing neural implants, sophisticated brain-computer interfaces, explicit methods of prolonging the healthy human life-span, regenerative medicine, mind uploading and so on.

The list of potentially transformative enhancement technologies is almost endless (see also Arnaldi, this volume).

The stalled debate on human nature

There is a broad spectrum of opinions concerning the desirability and the limitations of new technologies. Whereas some consider them to be nothing more than new and more sophisticated means for treating and preventing disease, for others these developments hold out the promise of exchanging the chance of the 'natural lottery' for free human choice, thus liberating us from the burdens of our biological determination. Enthusiastic voices speak of radically enhanced humans, indefinitely expanded life spans and the creation of superhuman artificial intelligence. The position of those eagerly embracing these prospects could be called *progressive technoeuphoric*, while *conservative technophobia* is characterized by the fear that our technological *hubris* might make us lose the essence of what it means to be human.

There are marked differences among countries concerning the readiness to embrace enhancement technologies and also a divide between the US and EU. The United States National Science Foundation-funded NBIC report (Roco and Bainbridge 2002), which has heavily influenced the debate around the world, had rather utopian overtones and a strong military focus. In a highly individualizing manner and with frequent references to social engineering, it highlighted the burgeoning potential of converging technologies to re-engineer human biology for improved performance. In order to outline a different approach to converging technologies and human enhancement the European Commission set up a High-Level Expert Group, 'Foresighting the New Technology Wave' which developed the concept of 'Converging Technologies for the European Knowledge Society' (CTEKS). Whereas the NBIC approach was seen as an attempt to 'engineer the mind and the body', CTEKS advocated 'engineering for the mind and the body' (Nordmann 2004). This initiative set out to adopt a demand-driven approach more in line with societal needs and expectations (Andler et al. 2008: 7) and laid more emphasis on involving a broader spectrum of disciplines beyond the natural sciences in discussion about converging technologies and their meaning for societies.

The question whether there is anything in human nature that is inherently valuable and deserving of 'protection' from technological manipulations has been one of the most intensely debated issues. On the one hand, bioconservative thinkers such as Leon Kass, Francis Fukuyama, Michael Sandel or Jürgen Habermas tend to argue that human nature imposes limitations on our techno-scientific pursuits and that it does serve as the basis of the kinds of political regimes we might come to have. They see enhancement technologies as threats that impinge upon human dignity (Kass 2002), upset the political order and create severe injustice (Fukuyama 2002, see also Arnaldi this volume), banish our appreciation of the given (Sandel 2007), or even transform the ethical self-understanding of our species and rob us of our autonomy (Habermas 2003). While most of these authors maintain that we must do everything in our power to treat disease and reduce suffering, they attempt to erect a boundary between purely curative and enhancing interventions. However, this demarcation needs to be drawn in an ever shifting terrain of contestations about normality and illness. A number of scholars have pointed out that the dynamics of medicalization – that is the redefinition of any condition, process, behaviour or characteristic as most properly belonging under medical jurisdiction as opposed to other descriptions – is of crucial relevance for discussions about enhancement (Conrad 2007). As ever more phenomena, ranging from (successful) aging, through childbirth to shyness are rendered intelligible in medical vocabularies, it becomes more and more difficult to maintain a clear line demarcating cures and improvements (Moynihan, Heath and Henry 2002).

Bioconservative arguments are met by technoprogressive thinkers like Nick Bostrom, John Harris, Julian Savulescu or Jonathan Glover, who argue that there is nothing prescriptive about the set of traits humans have developed through the chance process of evolution and point out that humans have always used technologies to shape themselves. Positioning their line of thought in the Enlightenment tradition, these thinkers emphatically embrace the prospect of emancipating humanity from the constraints of nature. One of the most outspoken advocates and proponents of the emancipatory potential of this technological self-transcendence is the transhumanist stream of thought, which 'affirms the possibility and desirability of fundamentally improving the human condition through applied reason, especially by developing and making widely available technologies to eliminate aging and to greatly enhance human intellectual, physical, and psychological capacities' (Humanity Plus 2003).

In her analysis of the debate on converging technologies, Arianna Ferrari (2008) has shown the inadequacy of an exclusive emphasis on the permissibility of modifying human nature. For her, this restricted focus leads to a severe impoverishment of the debate, which has come to circle around issues of minor relevance, driven more by science fiction scenarios than real world challenges. She also draws attention to the polyvalence of the nature-discourse, demonstrating how one side claims the normativity of human nature, thus arguing for the prohibition of its manipulation, whereas the other party employs the naturalness of striving for change, self-transcendence and reshaping ourselves. Ferrari also believes that the enhancement debate has become deadlocked as a result of the dominance of the 'human nature' issue and because both parties uncritically accept the arrival of radically transformative technologies. Consequently, the debate seems to be taking place between groups that either fear or embrace a future they all see as unavoidable. This leaves little room for the imagination of political actions other than preparing for the worst or moving the obstacles out of the way of technologically created bliss.

Whereas bioconservative thinkers argue in favour of a strict separation of human nature from a perceived impure technological impingement, others, such as transhumanists, jubilate the arrival of 'liberation biology' (Bailey 2005). Yet both positions seem to relate to technology as if it were something external and look at its application to humans either with fear or joyous expectation. However, instead of such a straightforward relationship, technology shapes, mediates and intimately permeates our lives in myriad ways. To draw on an especially succinct formulation: 'what we make and what (we think) we are co-evolve together' (Hayles 2006: 164). Our relation to technologies is thus not evident and univocal but highly ambiguous and ambivalent. Emerging technologies shall likely neither strip us of our 'humanness' nor make us absolute masters of our destiny, but entangle us in ever more complex relations.

The transformation of medicine and the biopolitics of enhancement

Throughout the second half of the twentieth century the remits of medicine extended 'beyond the dimension of illness and cure and into the management of normality itself' (Rose 1994: 67) and scholars have observed the rise of preventive, surveillance medicine, which seeks to constantly monitor health states to intervene as early as possible (Armstrong 1995). With the blurring of boundaries between prevention, treatment and enhancement, medicine may be undergoing an important transformation. Wiesing suggested that we are at the threshold of a new paradigm that involves a transition from *restitutio ad integrum* to *transformatio ad optimum,* that is, a shift from restoring integrity and health, to transforming it into the best (Wiesing 2009). This shift has been reaffirmed by Wolbring, who describes the gradual transhumanization of health and medicine, as a result of which 'normal' functioning and the merely 'healthy' body come to be viewed as 'limited, defective, impaired, and in need of constant improvement made possible by new technologies' (Wolbring 2010: 295). This transformation may be fruitfully analysed from a governmentality perspective and with recourse to Michel Foucault's notion of biopolitics (see also Pellizzoni and Ylönen this volume, Ferreira et al. this volume).

In a lecture at the College de France in 1976 Foucault discussed the concept of biopolitics, which he described as a technology of power that emerged in the 18th century. It is a form of power, characterized by an attempt to administer and monitor the life and vitality of populations. Complementary to disciplinary techniques that sought to control and condition the individual body at various sites, such as schools, prisons, factories, etc., this new form of biopower focused on the biological life of the human species qua species. In Foucault's words biopolitics

> [i]s a technology which aims to establish a sort of homeostasis, not by training individuals, but by achieving an overall equilibrium that protects the security of the whole from internal dangers. [...] Both technologies are obviously technologies of the body, but one is a technology in which the body is individualized as an

> organism endowed with capacities, while the other is a technology in which bodies are replaced by general biological processes. (Foucault 2003: 249)

He described the birth of biopolitics as the time when the management of life entered the realm of political calculations and when state control gradually took hold of the biological life of populations in the name of increasing and optimizing vitality (Foucault 2003: 243).

Such political techniques as statistics, demography and epidemiology allowed for the study and governing of populations, their rates of vitality, production and reproduction. Importantly, biopolitics entails a decreased preoccupation with battling large epidemics that used to haunt societies prior to the relative security brought about by the 18th century, and instead it focuses on endemics, or internal factors that threaten the health and productivity of nations (Foucault 2003). The centrality of purity, protection and productivity links biopolitical practices to the formation of nation states and the burgeoning of capitalism. As Foucault argued, this rearrangement of state rationality also meant that the protection of the life and gene pool of nations gained crucial relevance. Racism, the fear of biological decline and degeneration as well as classical eugenics, which aimed at improving the quality of the 'human stock', are all paradigmatic examples of this biopolitical logic. An important characteristic of biopolitics as described by Foucault follows in part from the attempt to optimize life processes; namely, governing through norms and concepts of normality, as opposed to a strictly 'juridico-discursive' model of power. Biopolitics operates by establishing standards and norms in self-conduct as well as in self-adjudication.

The present biopolitical landscape is in many ways different compared to even the 1970s, when Foucault was writing. The epistemic shift from the molar to the molecular, which has taken shape within biomedicine as a result of advances in molecular biology, such as the genetic engineering of bacteria, has recast our concepts of life itself and refocused the level of intervention into biological processes. This 'molecularization of life' (Rose 2007a: 11–15) has given rise to a new type of explanation for biological phenomena, which have gradually come to include the workings of the mind as well (Abi-Rached and Rose 2010), enabling new forms of manipulation.

> [W]hereas biochemistry and early molecular biology had attempted to produce an extra-cellular representation of the intracellular structures and processes, genetic engineering strove to do the reverse: to create […] the intracellular representation of an extracellular project. (Gottweis 1997: 65)

As the effort at understanding cellular mechanisms was supplanted by manipulation, life itself became malleable and began to appear as something to be improved upon. This transition brought with it a 'discourse of deficiency' (Gottweis 1997), which is an important facilitator of the burgeoning of enhancement medicine. As Gottweis argued, the emergence of novel biotechnologies coincided with deep recessions and post-war crises, which pointed toward a neoliberal re-organization of the global political economy. In this context the potential of novel technologies to intervene into biological processes, especially genetic engineering, was seen as a possible source of improving the quality of life, but also of creating social stability (Gottweis 1997: 66).

Also, Paul Rabinow's term 'biosociality' offers an expansion and elaboration of Foucault's concept of biopower. According to him, the Human Genome Initiative and related biotechnological innovations signaled a transformation of both the collectivizing biopolitical, population-oriented and the individualizing disciplinary, body-oriented poles that Foucault originally described. 'Biosociality' serves the purpose of delineating a post-disciplinary society in which the gulf between 'nature' and 'culture' had been dissolved and overcome, or at least reversed. His notion of a shift from sociobiology to biosociality refers to a change by virtue of which biology and biological categories cease to function as 'metaphor[s] for modern society', meaning that social projects will no longer set out to understand and organize the social along the lines of a perceived natural order. Instead, 'nature will be modeled on culture understood as practice. Nature will be known and remade through technique and will finally become artificial just as culture becomes natural' (Rabinow 1996: 92; see also Pellizzoni and Ylönen this volume). Though we may concede with Rabinow that 'nature's malleability offers an "invitation" to the artificial' (Rabinow 1996: 108), it is important to point out that this malleability may give rise to highly problematic interventions. As the dichotomy between nature and culture is dissolved and both become remade, we might witness the inscription of social norms and expectations at the level of the body, which would in effect constitute a re-naturalization of the social. In relation to enhancement, this means that 'the emancipation from the natural body and its inadequacies merges in the self-subordination to the social norm of body enhancement' (Wehling 2005:12).

Who we are is increasingly defined by what can be known about our biological make-up and by the proactive stance we take in relation to our natural endowments, either in the form of acting upon risks or by constantly striving for improvement. As our self-understanding comes to be shaped by biomedicine and by the actual and promised possibilities of biotechnologies, even our understanding of personhood is being interpreted 'by others, and by ourselves, in terms of our contemporary understandings of the possibilities and limits of our corporeality' (Rose 2007a: 76). As we become 'somatic individuals' (Novas and Rose 2000), we learn to think of our individuality in bodily terms, which means that the body comes to serve as the fundamental site of acting upon ourselves, by the means provided mostly by biomedicine. Biology becomes more 'open' to choice and we learn to act upon it and integrate knowledge about it into our lives. Simultaneously, we become responsible for the design we choose for our bodies (Negrin 2002). The possibility of intervention comes with inevitable responsibility (Harris 2007: 118). As we learn more about our biological makeup we are incited to take a proactive stance towards its management because

> the reorganization of many illnesses and pathologies along a genetic axis does not generate fatalism. On the contrary it creates an obligation to act in the present in relation to the potential futures that now come into view. (Rose 2007a: 107)

The obligation to act arises where the image of the prudent, self-governing individual confronts discourses of risk. Gaining knowledge and, to a far lesser extent, a possibility to intervene and act put individuals in a position where they need to reconsider some of their bonds to others, most notably to potential future kin. The pre-symptomatically ill, who have been identified to carry a predisposition for diseases with a genetic component, often find themselves entangled in the web of biomedicine and subjected to surveillance and preventive measures despite their perfectly normal present condition and maybe even without much certainty about whether they will ever develop a specific disease (Rose 2007a: 125). In fact, the reorganization of illnesses along a genetic axis also entails that fundamental notions, such as autonomy, are reinterpreted. In order to be autonomous and act prudently one *must* take genetic information into account, since failure to do so would not be seen as an individual choice but rather the demonstration of irresponsibility. The growing interest and research in the field of biomarkers for the prediction of human behaviour extends this logic beyond genetic information, so that all kinds of knowledge about what the body is like biologically are invested with more and more meaning and significance (Singh and Rose 2009). Developing Foucault's concepts of discipline and biopolitics further, Rose terms the type of politics that forms around the governance of somatic individuals *ethopolitics*.

> If discipline individualizes and normalizes, and biopower collectivizes and socializes, ethopolitics concerns itself with the self-techniques by which human beings should judge themselves and act upon themselves to make themselves better than they are. (Rose 2001: 18)

In a way, with the transhumanization of health and medicine everyone is becoming 'at risk' and needs to adopt a managerial attitude towards the body, constantly striving to improve it and overcome its limitations. Without direct influence, manipulation or state oppression individuals themselves internalize the norm of self-monitoring and the body becomes a crucial site for expressing and acting upon individuality. According to Foucault's definition, technologies of the self

> permit individuals to effect by their own means or with the help of others a certain number of operations on their own bodies and souls, thoughts, conduct, and way of being, so as to transform themselves in order to attain a certain state of happiness, purity, wisdom, perfection, or immortality. (Foucault 1988: 17)

The methods, skills and attitudes that individuals employ to shape themselves are inseparable from technologies of power, 'which determine the conduct of individuals and submit them to certain ends or domination' (Foucault 1988:

17). This intertwined relationship, their point of contact, is what Foucault terms 'governmentality' (see also Ferreira et al. this volume, Pellizzoni and Ylönen this volume). Biotechnological enhancement may be understood in this framework as a self-technology and a manifestation of neoliberal governmentality, in which the political goals of improved productivity are intertwined with self-technologies aimed at securing, optimizing and improving individual health and well-being.

In the following section I am going to elaborate on the idea of the neoliberal subject and its relation to biotechnological enhancement.

The neoliberal subject

Luc Boltanski and Eve Chiapello distinguish three different stages of capitalism, each of which employed different strategies of mobilization to ground their legitimacy. The authors argue that each stage provoked specific forms of criticism to which capitalism responded by transforming itself so as to incorporate certain points of the critique. The promise of emancipation has been a central element of capitalism ever since the 19th century and it presented itself as a form of liberation from the bonds of traditional society, which crippled individual efforts of self-realization.

> In contrast to societies where people are assigned a status that is practically impossible to alter in the course of their lives [...] capitalism supposedly offers the possibility of voluntary deracination, which is protected by the importance it accords to the legal device of the contract. For a contract, unlike status, can be established for a fixed term, and does not engage the whole of a person, but stipulates the particular respect in which the person is bound by their promise in their relationship with someone else. (Boltanski and Chiapello 2007: 426)

Therefore, part of the claim to legitimacy of the first spirit of capitalism in the 19th century was its promise of autonomy, the liberation from local structures and the promise of participation in progress. The critique of this era, which Boltanski and Chiapello label a 'social critique', claimed that the dissolution of traditional bonds and the universal competition of all against all had not brought liberation but rather resulted in 'enslavement to factory discipline, and meagre pay [which] no longer allow for the realization of a properly human existence' (Boltanski and Chiapello 2007: 427). A second line of critique, along Marxian lines, held that for its continued functioning capitalism depended on the unlimited flow of goods, yet this unchecked and unlimited production was predicated on the creation of an endlessly consuming subject with insatiable desires. This radically egoistic individual, set in the stage of a laissez-faire economy was seen as destined to bring social disintegration (Boltanski and Chiapello 2007: 428).

According to Boltanski and Chiapello, the second spirit of capitalism, roughly from the 1930s to the 1970s responded to these challenges and brought the promise of welfare and security by implementing strategies 'for stabilizing and coordinating actions, strengthening institutional boundaries, and planning and bureaucratization' (Boltanski and Chiapello 2007: 428). This 'welfare era' might be seen as a period that had provided a much more tangible form of liberation. Yet critique directed against this spirit of capitalism raised the contention that security had been bought at the price of sacrificing individuality and difference. Standardization and bureaucratization were denounced as forms of mass producing not only objects but also human beings, who are thus reduced to minor, meaningless and interchangeable cogs within the great, self-serving machinery. *Brave New World*, being a fearful depiction of a society that marries Fordist assembly-line production with the idea of eugenics, is an exceptionally strong formulation of this concern.

In response, the transformation of capitalism since the 1980s involved attributing heightened relevance to individual actions and autonomy. However, the possibility of self-realization demanded giving up on claims to a stable and calculable future, and a corresponding decrease in state provisions and welfare. This process was most emphatically expressed by Margaret Thatcher's famous statement, claiming: there's no such thing as society, only individual men and women who must first and foremost care for themselves (Thatcher 1987).

The emergence of neoliberal capitalism in 'advanced liberal societies' has given rise to an *enterprise culture* and led to the emergence of what Bröckling (2007) calls the *entrepreneurial self*, which is an image of the self as autonomous and conscious – a choosing self, who pursues its own life-plans according to its own values and priorities.

> The self is to be a subjective being, it is to aspire to autonomy, it is to strive for personal fulfilment in its earthly life, it is to interpret its reality and destiny as a matter of individual responsibility, it is to find meaning in existence by shaping its life through acts of choice. (Rose 1996: 151)

The current 'flexible' capitalism thus incites individuals to initiate as many 'projects' as possible and foregrounds the importance and value of having 'no strings attached'. In fact, the level of self-fulfilment an individual has achieved has been 'elevated to the status of an evaluative criterion' (Boltanski and Chiapello 2007: 429) such that living an active, rich, productive and fulfilling life are positioned as moral values. As Nikolas Rose puts it, 'contemporary individuals are incited to live as if making a project of themselves, [...] to develop a "style" of living that will maximize the worth of their existence to themselves' (Rose 1996: 157).

During the 1980s and 1990s the principle of self-governance and individual responsibility gained prevalence in relation to issues of health as well. As Gottweis notes, the 'idea of the managing of the self is reflected in a multitude of technical and organizational novelties within healthcare, in which managed care is the most important and most paradigmatic example' (Gottweis 2005: 32). Responsibility and individualized project thinking extend to the level of managing our biological constitution, which is becoming increasingly relevant and thematized along the lines of a valuable asset. Yet, while neoliberal narratives constantly reinforce the role of personal choice, individual autonomy, responsibility and self-governance, we can also observe the dislocation of the individual from her privileged position. The following points may serve to underline this observation.

Martin Weiss (2009) argues that recent biotechnological developments have dismantled our previous understanding of 'human nature' as something solid and unchangeable and have made it fundamentally malleable. Human nature now stands for a wide array of manipulable biological traits, susceptibilities, neurotransmitter levels and so forth. As a result, the individual is de-constructed, or de-composed and finds itself dissolved in various types of biological data, preserved in anonymous biobanks. Decorporalization can thus be considered a crucial element of current biopolitics as the materiality of the body is dissolved into informational entities and statistical probabilities (see Gottweis 2005: 33). The dissolution of the subject is complemented by the dissolution of classical forms of state sovereignty, as biopolitical grand projects give way to a dispersed network of performative discourses that simultaneously construct and dispel the idea of an autonomous individual (Weiss 2009: 51). We are incited to act responsibly in the management of vitality; yet the principle shaping such 'prudent' action is increasingly informed by communitarian principles and the 'good' of society. As Knoppers and Chadwick have argued, the principle of solidarity and a 'communitarian turn' are central developments in recent bioethics (Knoppers and Chadwick 2005). Weiss concludes that self-governance and heteronomy are inseparably intertwined and raises the important question concerning the nature of a community comprised of 'dissolved' individuals (Weiss 2009: 52).

Following Lessenich, contemporary advanced liberal societies may be described as 'active' or 'neosocial' societies which guide citizens toward the

> adoption of 'pro-active' behavior, understood as planned, purposive, and prudential action. Pro-active behavior means calculating and (on that basis) taking risk(s), adopting an entrepreneurial stance toward life, relying on self-management and self-control. (Lessenich 2010: 312)

In the active society individual responsibility includes not only oneself but, rather, it encompasses responsibility for the whole, for society as well. 'Socialized selves', in Lessenich's terminology, are obliged to act prudently in order to ensure the benefit of society through their own individual actions; thus, the 'common good' becomes the assumed maxim of responsible action (Lessenich 2010:306). According to Maasen, Sutter and Duttweiler,

> society is now prime reference of sociality and evaluates individual activities according to their degree of sociality. This requires the individual's capacity to monitor and control themself – for the benefit of themself and society.

> [...] Being neosocial is thus tantamount to individuals that flexibly govern themselves and others by way of socially accepted means. (Maasen, Sutter and Duttweiler 2007: 28)

The vision of self-fashioning, self-monitoring subjects aligns well with an individualized enhancement regime. For example, in an attempt to differentiate the 'old eugenics' from current practices Julian Savulescu argued:

> What was objectionable about the eugenics movement, besides its shoddy scientific basis, was that it involved the imposition of a State vision for a healthy population and aimed to achieve this through coercion. The eugenics movement was not aimed at what was good for individuals, but rather what benefited society. Modern eugenics in the form of testing for disorders, such as Down syndrome, occurs very commonly but is acceptable because it is voluntary, gives couples a choice over what kind of child to have, and enables them to have a child with the greatest opportunity for a good life. (Savulescu 2009: 526)

However, this distinctively neoliberal reasoning can be questioned from the perspective of empirical research, which has shown that real-life decision making in the face of reproductive choices is far more constrained and limited than theories seem to suggest. Furthermore, an increase in possible choices often negatively affects people's capacity to make decisions (Rapp 1999). These constraints can be 'due to structures, due to medical professionals' desire to help, due to pressures in society and due to technology as such' (Zeiler 2004: 179). Individual choice itself is articulated within an ethopolitical framework and guided by the neosocial value of not only individual responsibility, but the good of society as well.

The rhetoric of self-management and self-enhancement is inseparably linked to discourses on pre-emption, prevention, anxiety and security. The next section introduces the notion of the 'anxious self', which serves to exemplify how self-practices and experiences of oneself are reshaped to live up to the task of properly, prudently and responsibly managing vitality.

The anxious self

There are many ways of understanding anxiety. On the one hand, it is an affective state of unease, worry and restlessness that seems to lack any specific directedness and which is probably common to most of us. This feeling of 'homelessness' has also played a central role in the existentialist philosophies of Søren Kierkegaard ([1844] 1980) and Martin Heidegger ([1927] 1962). Here, the free floating, ungraspable and indefinable feeling of anxiety constitutes the fundamental condition of being human. It is through the torments of anxiety that the depths of life are revealed. For these thinkers anxiety is essential and productive for it is the precondition of human freedom and deep reflection, wherefore anxiety needs to be faced and lived rather than evaded.

On a slightly different note, anxiety has been associated with the discontent brought about by modernity itself. Processes of modernization loosen traditional bonds, introduce relativism and seem incapable of providing a framework of meaning for the lives of individuals beyond a bleak vision of progress, which itself has been largely shattered by the cataclysms of the twentieth century. It is a recurrent theme in a number of writers since the nineteenth century, both in the US and in Europe, to condemn their age as one of anxiety and uncertainty (Wilkinson 2001). Anxiety has been considered an expression of the individual's struggle in a world she perceives to be 'wrong' in a profound sense. The idea that the very form of life characteristic of modern societies engenders feelings of insecurity has a fairly long history such that by the end of the last century it had become something of a commonplace.

How should we understand anxiety as a technology of the self in this sense? The current biopolitical landscape brings forth a certain norm, namely the norm of constant worry and rumination over one's biology. It is important to note that I am not arguing for the case that individuals have necessarily become more anxious in the psychological sense.[1] I also believe the dynamics at work here are not like everyday fear of disease or illness. Rather, it seems to me to be the case that certain analogies exist between states of anxiety and forms of self-governance.

One such analogy is the increased concern with which we deal with our biological existence. The internalization of the norm of restless concern has almost become an integral part of our notion of responsibility. The term 'anxious subject' is in my view an adequate one because the type of worry that current discourses of risk instill is not necessarily targeted at anything specific. Being concerned and worried about our bodily state in general becomes the norm. The increasing role that our biological characteristics and bodily well-being play in making sense of our identity simultaneously destabilizes the lived experience of the body. As Martin Weiss argues, previous discourses of genetic determinism solidified a certain biological destiny, but they also allowed the affected person to develop an attitude of acceptance and learn to live with the given. Contrary to this, discourses of susceptibility seem to construe the whole of an individual's biological constitution as a source of risk, as something that poses potential threats unless carefully supervised (Weiss 2009).

Of course, in a certain sense the body has always been a source of great anxiety because at some fundamental level it is beyond the individual's rational control and serves as the most vivid reminder of our vulnerable, fragile and ephemeral nature. However, the phenomenology, the lived experience of the body, has thus far remained mostly uncontested, whereas now categories such as 'at risk' or

---

1 Even though the incidence of anxiety disorders seem to be on the rise, the cause of this is not easy to determine, and may be due to the medicalization of negative affects instead of actual rise in incidence. See also Horwitz and Wakefield (2007).

'pre-symptomatically ill' dislocate, or at least contest, first-person experience by claiming to speak the language of scientific objectivity.

Experiences of derealization and depersonalization – during which one's body or surroundings are experienced as frighteningly alien – often form symptoms of anxiety disorders (Simeon et al. 2003). This sense of derealization seems analogous to the kind of dissociation that risk thinking involves. The way severely anxious states may involve that one feels estranged from the world or one's body, so does being 'at risk' dissociate experience from 'fact.' To belong to the group of the pre-symptomatically ill entails that one's first person account of herself and her health contradicts that of medical discourse, which claims to speak in the name of objective biological reality. In essence the person is alienated from her own lived experience. Furthermore, probabilistic accounts of risks are generally perceived to be alien from everyday thinking and therefore very difficult to grasp and integrate, while putting forth a strong claim to authority (Lock 2007).

The ever broadening spectrum of medicalization is also a strong catalyst of this process. As more and more phenomena come to be viewed as belonging under medical jurisdiction – such as birth, (successful) ageing, reproduction, nutrition, beauty, intellectual fitness, emotional life, etc. – more and more areas of life require our prudent, active engagement and careful concern. We also become more dependent on medical vocabularies to make sense of our own experiences. This, in effect, leads to a form of constant self-monitoring, which is also quite akin to that of anxious states. To exaggerate the situation slightly: each and every sign of the body, signs of its functioning, ageing and change, become invested with great meaning and may be interpreted as potential (medical) problems. Interestingly, this state of over-concernedness seems analogous to panic disorders, which are also located within the spectrum of anxiety disorders. In such cases perfectly healthy individuals experience sudden and short lasting states of severe restlessness and tension, which manifests itself in the form of physical symptoms such as the racing of the heart, shivers, chills, sweating, dyspnea, dizziness, and so on. These experiences then lead to fear of other attacks and induce persistent worry, and fear of consequences such as dying or going mad (APA 2000).

An interesting demonstration of the way this regime of anxious self-concern may become normalized is provided by Carmen Baumeler's analysis of 'affective computing'. Such, as of yet hypothetical, systems are wearable computing devices that monitor stress related physiological changes in order to help prevent cardiovascular disease. The system is also linked to a center where an individual health expert monitors the values and via video link gives advice on how to manage distress and negative emotions. In an example, the user devotes a considerable amount of attention to the handling of her stress levels and checks in to see her values about five times a day. According to Baumeler 'this application demonstrates, [that] users are supposed to manage stress themselves and, therefore, stay healthy and productive' (Baumeler 2008: 188). The author intends this example to show how the individualization of emotion management is linked to the production of the ideal 'flexible worker'. The objectifying gaze directed at one's own body, this constant rumination, the drive to check, recheck and double-check to see whether everything is OK according to some scientifically established criterion, is also the hallmark of anxious self-concern. This example seems to be the paradigmatic expression of the responsible/anxious subject and the establishment of a norm.

Pre-emptive enhancement

Thomas Lemke describes a generalized dispositive of risk, which is more and more dissociated from any concrete or given danger, whereby a form of restless uncertainty in effect becomes the normal condition. He uses the examples of terrorism, and biological risk, which threaten to dissolve the health and integrity of the collective and the individual body respectively (Lemke 2004: 16). This kind of risk-thinking is especially prevalent in discourses on genetics. According to Lemke, the scientifically outdated idea of genetic determinism is still very much present and is slowly complemented by a post-genomic view. With this transition the idea of genes as an underlying layer causally determining the surface is supplanted by an understanding which accords genes themselves a different and somewhat more modest role (Keller 2000). Except in the cases of single gene disorders this entails a shift in emphasis from causality to susceptibility and probability. The focus on susceptibility, screening and prevention seems to promise more security and the chance to intervene or adjust to the knowledge gained about individual predispositions, yet, according to Lemke, the result is quite the contrary (Lemke 2000).

Besides the prevention of individual risks to counter susceptibilities, the identification of people who present certain threats to broader society is also prevalent. Calls for the creation of population-wide forensic genetic databases (Townsend and Ashtana 2008), programs to screen for potentially dangerous personality disorders and the introduction of the Indeterminate Public Protection Sentence in the UK in 2003[2] illustrate this well. Surveillance already reaches peak levels with the aim of achieving total information awareness[3] of all suspicious activities and persons.

> Since we cannot screen out 'them' who are threatening the status quo, measures are increasingly directed towards the domestic spaces. This materializes among others in the form of mass data collection. In a state of fear prevention became the magic word, for the sake of which civil rights and procedural guarantees are given up. (Bárd 2008: 100)

2 See Prisoners' Advice Service - Information Sheet. Available at: www.prisonersadvice.org.uk/DOCS/INFORMATION/IPP.pdf [accessed: 19 July 2011].

3 See http://www.fas.org/irp/agency/dod/poindexter.html [accessed: 23 December 2011].

The technique of filtering through mass databases in search for associations that suggest suspicious activities and then acting upon such algorithmic probabilities has gained prevalence since the events of 9/11 and has contributed to the securitization of everyday life (Amoore 2009), and technologies for extending surveillance to human biology are also being developed. Melinda Cooper describes the biological turn in the war on terror, which 'conflates public health, biomedicine, and war under the sign of the emerging threat' (Cooper 2008:75). MALINTENT, a system developed by the United States Department of Homeland Security to scan biomarkers, such as blood pressure, heart rate, breath-rate and non-verbal cues in order to identify harmful intentions, is one pertinent example (Barrie 2010). MALINTENT is planned to be employed at airports but can easily be set up at any location to screen out harmful individuals. The analogy to an Orwellian thought-police is almost too plain.

There is increased interest in forms of 'soft surveillance' as well, which are non-intrusive and can remain ubiquitous (Marx 2006). In such attempts the autonomous, choosing self is truly dissolved in biological markers which are then reconstructed along the binary logic of 'threatening/not-threatening'. It seems that the – however conceived – interests of the community evidently trump the privacy rights of 'dissolved' individuals.

But how does all this relate to human enhancement?

Whereas earlier attempts to increase the quality of the human stock were centrally orchestrated and directed at entire populations, scholars have argued that a new and individualized enhancement regime is on the rise since the 1980s. Though today's political interest in public health is usually not framed in terms of population unfitness or geopolitical struggles between nation states, we can find examples that evoke versions of these classical eugenic arguments.

The question of geopolitical competitiveness is raised frequently in relation to countries such as China, which is portrayed as the most likely candidate for pursuing 'authoritarian transhumanism' (ETAG 2006) and where scientists may be far less constrained by ethical considerations (PCB 2003:40). Some even speak of a eugenics 'arms race' with China (Swedin 2006). In this context, the phrase 'China will do it anyway' has almost become a commonplace in discussions about enhancement and it frames the question as a matter of competitiveness among countries (Coenen et al. 2009).

Yet, it also seems that the ever expanding scope of securitization and the protection of the 'good of society' come to encompass human biological functioning as well. The transhumanization of medicine and the desire to ameliorate the human condition link up with discourses on risks and susceptibilities so that a number of human 'deficiencies' come to be construed as immense threats. As a result, the biological constitution of the human being comes to be seen as inherently lacking in a number of respects and turns into an object of biotechnological improvement and optimization. When this is rendered in the framework of security, enhancement becomes a form of prevention aimed at managing the risks of being human.

Persson and Savulescu (2011) have argued that humans may be 'unfit' for the future because our evolutionary heritage makes us ill-adapted to life under circumstances of plurality within mass societies, in which immense destructive technological power can easily get hijacked and wreak havoc upon millions. They highlight the moral shortcomings of our species, as evidenced by the limited scope of the moral circle and our failures to act fairly and responsibly in a globalized world. Of special concern to them are security threats posed by terrorists and the danger of sophisticated technologies falling in the wrong hands. As possible and complimentary solutions they consider restricting liberal neutrality, strengthening moral education to achieve more equality, and greater infringements of privacy and surveillance, as well as applying enhancement technologies to remedy the 'misfit' between our biology and our way of life.

On this account enhancement is no longer proposed as an option of individual choice; it is no longer an issue of morphological freedom, distributive justice, or a question of emancipating humanity from nature's confines. It becomes a preventive measure. Human deficiencies are presented as potential existential threats, which call for extraordinary procedures and a radical extension of what Rose (2010) described as 'screen and intervene'. This – as yet hypothetical – act can be interpreted as the securitization of human nature, where liberal principles are suspended under the banner of preserving liberalism (Opitz 2010).

Conclusion

As medicine becomes increasingly concerned with optimizing and improving functioning beyond what is considered healthy at any given moment in time, the ideal of self-fashioning and self-enhancement gains prevalence. I have argued that the neoliberal rhetoric of personal choice in shaping ourselves is complemented by growing anxiety about the body and is guided by the always presupposed value of promoting the good of society through prudent individual choices and action. I tried to highlight the constant expansion of preventive and pre-emptive measures to counter risks and argued that we may currently be witnessing the gradual transformation of the notion of enhancement technologies into such preventive and pre-emptive interventions. At the point where the culture of danger renders human 'biological deficiencies' as matters that necessitate the suspension of liberal values, the security *dispositif* runs the risk of dissolving 'its constitutive *aporia* and [inverting] it into a sovereign machine' (Opitz 2010: 100).

The question we are left with is: how can we find new and creative ways of understanding human enhancement, disentangled from discourses on neoliberal productivity and securitization?

References

Abi-Rached, J. and Rose, N. 2010. The birth of the neuromolecular gaze. *History of the Human Sciences*, Special issue on the new brain sciences, 23(1), 11–36.

Andler, D., Barthelmé, S., Beckert, B., Blümel, C., Coenen, C., Fleischer, T., Friedewald, M., Quendt, C., Rader, M., Simakova, E. and Woolgar, S. 2008. *Converging Technologies and Their Impact on the Social Sciences and Humanities*. Final Report of the CONTECS Project. European Commission 6th Framework Programme. [Online]. Available at: http://www.contecs.fraunhofer.de [accessed: 27 December 2011].

APA. 2000. *Diagnostic and Statistical Manual of Mental Disorders* (4th ed., text rev.). Washington, DC: American Psychiatric Association.

Amoore, L. 2009. Algorithmic war: Everyday geographies of the war on terror. *Antipode*, 41(1), 49–69.

Armstrong, D. 1995. The rise of surveillance medicine. *Sociology of Health & Illness*, 17(3), 393–404.

Bailey, R. 2005. *Liberation Biology: The Scientific and Moral Case for the Biotech Revolution.* New York: Prometheus Books.

Bárd, P. 2008. The principle of availability and the exchange of DNA profiles among Member States: The development of forensic sciences or that of a European Panopticon. *Ügyészek Lapja* [Prosecutors' Journal], Különszám [Special Edition], Budapest, 99–108.

Barrie, A. 2010. *Homeland Security Detects Terrorist Threats by Reading Your Mind.* Available at: http://www.foxnews.com/story/0,2933,426485,00.html [accessed: 19 July 2011].

Baumeler, C. 2008. Technologies of the emotional self: Affective computing and the 'enhanced second skin' for flexible employees, in *Sexualized Brains. Scientific Modeling of Emotional Intelligence from a Cultural Perspective*, edited by N.C. Karafyllis and G. Ulshöfer. Cambridge, MA: MIT Press, 179–90.

Boltanski, L. and Chiapello, E. 2007. *The New Spirit of Capitalism*. London: Verso.

Bröckling, U. 2007. *Das Unternehmerische Selbst*. Frankfurt am Main: Suhrkamp.

Coenen, C., Schuijff, M., Smits, M., Klaasen, P., Hennen, L., Rader, M. and Wolbring, G. 2009. *Human Enhancement*. Brussels: European Parliament, Science and Technology Options Assessment (STOA). Available at: http://www.itas.fzk.de/deu/lit/2009/coua09a.pdf [accessed: 23 December 2011].

Conrad, P. 2007. *The Medicalization of Society*. Baltimore: Johns Hopkins University Press.

Cooper, M. 2008. *Life as Surplus*. Seattle: University of Washington Press.

ETAG. 2006. *Technology Assessment on Converging Technologies. Report for the European Parliament.* Brussels: European Parliament, European Technology Assessment Group. Available at: http://www.itas.fzk.de/eng/etag/document/beua06a.pdf [accessed: 23 December 2011].

Ferrari, A. 2008. Is it all about human nature? Ethical challenges of converging technologies beyond a polarized debate. *Innovation: The European Journal of Social Science Research*, 21(1), 1–24.

Foucault, M. 1988. Technologies of the self, in *Technologies of the Self: A Seminar with Michel Foucault*, edited by L.H. Martin, H. Gutman and P.H. Hutton. Amherst, MA: University of Massachusetts Press, 16–49.

Foucault, M. 2003. *Society Must Be Defended*. New York: Picador.

Fukuyama, F. 2002. *Our Posthuman Future*. New York: Picador.

Gottweis, H. 1997. Genetic engineering, discourses of deficiency, and the new politics of population, in *Changing Life. Genomes, Ecologies, Bodies, Commodities*, edited by P. Taylor, S. Halfon and P. Edwards. Minneapolis: University of Minnesota Press, 56–84.

Gottweis, H. 2005. Biobanks in action: New strategies in the governance of life, in *Biobanks – Governance in Comparative Perspective*, edited by H. Gottweis and A. Petersen. Abingdon: Routledge, 22–38.

Habermas, J. 2003. *The Future of Human Nature*. Cambridge: Polity Press.

Harris J. 2007. *Enhancing Evolution: The Ethical Case for Making Better People*. Princeton, NJ: Princeton University Press.

Hayles, K. 2006. Unfinished work: From cyborg to cognisphere. *Theory Culture & Society*, 23(7–8), 159–66.

Heidegger, M. [1927] 1962. *Being and Time*. Oxford: Blackwell.

Horwitz, A.W. and Wakefield, J.C. 2007. *The Loss of Sadness – How Psychiatry Transformed Normal Sorrow into Depressive Disorder*. New York: Oxford University Press.

Humanity Plus. 2003. *The Transhumanist FAQ: A General Introduction—Version 2.1*. [Online]. Available at: http://humanityplus.org/learn/philosophy/faq #answer 19 [accessed: 20 December 2011].

Huxley, A. 1998. *Brave New World*. New York: Perennial Classics.

Juengst, E. 1998. 'The meaning of enhancement', in *Enhancing Human Traits: Ethical and Social Implications*, edited by E. Parens. Washington, DC: Georgetown University Press, 29–47.

Kass, L. 2002. *Life, Liberty and the Defense of Dignity: The Challenge for Bioethics*. San Francisco: Encounter Books.

Keller, E.F. 2000. *The Century of the Gene*. Cambridge, MA: Harvard University Press.

Kierkegaard, S. [1844] 1980. The concept of anxiety, in *Kierkegaard's Writings*, Vol. 8. Princeton, NJ. Princeton University Press.

Knoppers, B.M. and Chadwick, R. 2005. Human genetic research: Emerging trends in ethics. *Nature Reviews: Genetics*, 6(1), 75–9.

Lemke, T. 2000. Regierung der Risiken. Von der Eugenik zur genetischen Gouvernementalität, in *Gouvernementalität der Gegenwart*, edited by U. Bröckling, S. Krasmann and T. Lemke. Frankfurt am Main: Suhrkamp.

Lemke, T. 2004. *Die politische Ökonomie des Lebens – Biopolitik und Rassismus bei Michel Foucault und Giorgio Agamben*. Available at: http://www.

thomaslemkeweb.de/publikationen/Die%20politische%20%D6konomie%20 des%20Lebens%20II.pdf [accessed: 20 December 2011].

Lessenich, S. 2010. Constructing the socialized self: Mobilization and control in the 'Active Society', in *Governmentality: Current Issues and Future Challenges*, edited by U. Bröckling, S. Krasmann and T. Lemke. New York: Routledge, 304–20.

Lock, M. 2007. The future is now: Locating biomarkers for dementia, in *Biomedicine as Culture*, edited by R.V. Burri and J. Dumit. New York: Routledge, 61–86.

Maasen, S., Sutter, B. and Duttweiler, S. 2007. Self-help: The making of neosocial selves in neoliberal society, in *On Willing Selves. Neoliberal Politics and the Challenge of Neuroscience*, edited by S. Maasen and B. Sutter. New York: Palgrave Macmillan, 25–52.

Marx, G. 2006. Soft surveillance: The growth of mandatory volunteerism in collecting personal information— 'Hey Buddy Can You Spare a DNA?', in *Surveillance and Security,* edited by T. Monahan. New York: Routledge, 37–56.

Moynihan, R., Heath, I. and Henry, D. 2002. Selling sickness: The pharmaceutical industry and disease mongering. *British Medical Journal*, 324(7342), 886–90.

Negrin, L. 2002. Cosmetic surgery and the eclipse of identity. *Body & Society*, 8(4), 21–42.

Nordmann, A. 2004. (Ed.) *Converging Technologies – Shaping the Future of European Societies.* High Level Expert Group 'Foresighting the New Technology Wave'. Luxembourg: Office for Official Publications of the European Communities. Available at: http://ec.europa.eu/research/social-sciences/pdf/ntw-report-alfred-nordmann_en.pdf [accessed: 20 December 2011].

Novas, C. and Rose, N. 2000. Genetic risk and the birth of the somatic individual. *Economy and Society*, 29(4), 485–513.

O'Connor, A. 2004. Wakefulness finds a powerful ally. *The New York Times*, 29 June.

Opitz, S. 2010. Government unlimited: The security dispositif of illiberal governmentality, in *Governmentality: Current Issues and Future Challenges*, edited by U. Bröckling, S. Krasmann and T. Lemke. New York: Routledge, 93–114.

Persson, I. and Savulescu, J. 2011. Unfit for the future? Human nature, scientific progress, and the need for moral enhancement, in *Enhancing Human Capacities*, edited by J. Savulescu, R.T. Meulen, and G. Kahane. Oxford: Wiley-Blackwell, 486–500.

PCB. 2003. *Beyond Therapy: Biotechnology and the Pursuit of Happiness.* Washington, DC: President's Council on Bioethics. Available at: http://bioethics.georgetown.edu/pcbe/reports/beyondtherapy/beyond_therapy_final_webcorrected.pdf [accessed: 20 December 2011].

Rabinow, P. 1996. Artificiality and enlightenment: From sociobiology to biosociality, in *Essays on the Anthropology of Reason*. Princeton, NJ: Princeton University Press, 91–111.

Rapp, R. 1999. *Testing Women, Testing the Foetus: The Social impact of Amniocentesis in America*. New York: Routledge.

Repantis, D., Schlattman, P., Laisney O. and Heuser, I. 2010. Modafinil and methylphenidate for neuroenhancement in healthy individuals: A systematic review. *Pharmacological Research*, 62(3), 187–206.

Roco, M. and Bainbridge, W. 2002. (Eds.) *Converging Technologies for Improving Human Performance*. Arlington, VI: National Science Foundation. Available at: http://www.wtec.org/ConvergingTechnologies/1/NBIC_report.pdf [accessed: 20 December 2011].

Rose, N. 1994. Medicine, history and the present, in *Reassessing Foucault. Power, Medicine and the Body*, edited by C. Jones and R. Porter. New York: Routledge, 48–72.

Rose, N. 1996. *Inventing Our Selves*. Cambridge: Cambridge University Press.

Rose, N. 2001. The politics of life itself. *Theory Culture & Society*, 18(6), 1–30.

Rose, N. 2003. Neurochemical selves. *Society*, 41(1), 46–59.

Rose, N. 2007a. *The Politics of Life Itself: Biomedicine, Power, and Subjectivity in the Twenty-First Century*. Princeton, NJ: Princeton University Press.

Rose, N. 2007b. Beyond medicalization. *The Lancet*, 369 (9562), 700–701.

Rose, N. 2010. 'Screen and intervene': Governing risky brains. *History of the Human Sciences*, 23(1), 79–105.

Sandel, M. 2007. *The Case Against Perfection.* Cambridge, MA: Harvard University Press.

Savulescu, J. 2009. Genetic interventions and the ethics of enhancement of human beings, in *The Oxford Handbook of Bioethics*, edited by B. Steinbock. Oxford: Oxford University Press, 516–35.

Simeon, D., Knutelska, M., Nelson, D. and Guralnik, O. 2003. Feeling unreal: A depersonalization disorder update of 117 cases. *Journal of Clinical Psychiatry*, 64(9), 990–97.

Singh, I. and Rose, N. 2009. Biomarkers in psychiatry. *Nature*, 460 (7252), 202–7.

Swedin, E.G. 2006. Designing babies: A eugenics race with China? *The Futurist*, 40(3), 18–21.

Thatcher, M. 1987. Interview for *Woman's Own*. Available at: http://www.margaretthatcher.org/speeches/displaydocument.asp?docid=106689[accessed: 19 July 2011].

Townsend M. and Ashtana, A. 2008. Put young children on DNA list, urge police. *The Guardian* [Online, 9 August] Available at: http://www.guardian.co.uk/society/2008/mar/16/youthjustice.children [accessed: 19 July 2011].

Wehling, P. 2005. Social inequalities beyond the modern nature-society divide? *Science, Technology & Innovation Studies*, 1(1), 3–15. [Online]. Available at: http://www.sti-studies.de/ojs/index.php/sti/article/view/68 [accessed: 27 December 2011].

Weiss, M. 2009. Die Auflösung der menschlichen Natur, in *Bios und Zoë*, edited by M. Weiss. Frankfurt am Main: Suhrkamp, 34–54.

Wiesing, U. 2009. The History of medical enhancement: From restitutio ad integrum to transformatio ad optimum? in *Medical Enhancement and Posthumanity*, edited by B. Gordijn and R. Chadwick. *The International Library of Ethics, Law and Technology*, Vol. 2, New York: Springer, 25–37.

Wilkinson, I. 2001. *Anxiety in a Risk Society*. London: Routledge.

Wolbring, G. 2010. Nanotechnology and the transhumanization of health, medicine, and rehabilitation, in *Controversies in Science and Technology Volume 3: From Evolution to Energy*, edited by D.L. Kleinmann, J. Delborne, K.A. Cloud-Hansen, and J. Handelsman. New Rochelle, NY: Mary Ann Liebert Publishers, 290–303.

Zeiler, K. 2004. Reproductive autonomous choice – A cherished illusion? Reproductive autonomy examined in the context of preimplantation genetic diagnosis. *Medicine, Health Care and Philosophy*, 7(2), 175–83.

Chapter 6

The question of citizenship and freedom in the psychiatric reform process: A possible presence of neoliberal governance practices

Arthur Arruda Leal Ferreira, Karina Lopes Padilha,
Míriam Starosky and Rodrigo Costa Nascimento

Introduction

There exists in Brazil a recent and interesting movement that changes psychiatry as theory and practice in a very complex way. It is sometimes called the anti-asylum movement (in general referring to the asylum's more extreme practices), but this movement is better known as 'psychiatric reform'. This movement is related to a great number of reformist efforts, such as the British one (anti-psychiatry), and especially the Italian reformist movement (democratic psychiatry). All these movements tried to put into question the notion of the asylum as the cornerstone of the treatment process, as well as the asymmetry between psychiatrists and patients, introducing political concepts in the analysis of psychiatric phenomena and devices.

Tracing its beginnings to the end of the 1970s, the Brazilian Psychiatric Reform opened the way to a few concepts that were, until then, considered incompatible with psychiatry, such as 'freedom', 'citizenship', and 'human rights'. At the same time it opened new places to the so-called 'mentally ill': the external space of the cities, assemblies, free work, consumption, and the responsibility of self-management. And finally new characters entered the scene in this old monologue with psychiatry: psychologists, sociologists, psychoanalysts, and social workers appeared as protagonists. These innovations were not abstract; they were supported not only by a great number of 'open-door institutions', but also by laws and governmental policies. In two decades psychiatry underwent a major change: the asylum disappeared as the cornerstone of therapeutic intervention and, at the same time, the psychiatrist lost his or her near omnipotence regarding madness.

In general, this process is celebrated as a kind of revolution that freed madness from the chains of old psychiatry. As with the major freedom movements, a special day is celebrated: May 18,[1] together with great narratives that confirm its triumph against the old and conservative forces. In this sense, a great part of the historiography related to this subject possesses this sense of an epic hagiographic narrative. Perhaps this historiographical celebration was necessary to create and recreate the dispositions to sustain the battle against the conservative forces in psychiatry, just in the same way as Foucault (1984) interpreted Kant's reference to the importance of the French Revolution.

Nevertheless, we think that history and STS can support this struggle with many more interesting and powerful resources. First, they do this by considering the 'psychiatric reform' as a much more complex technoscientific process related to a variety of political and governmental practices. Second, they open space to problematizations and to the analysis of new dangers present in these reformist processes (considering other factors beside the asylum institutions).

Amidst the activists of the 'psychiatric reform', it is very frequent to consider some critics of the reform process and devices as a manifestation of the old and conservative psychiatry. And frequently this supposition is correct; sometimes some conservative critics appear in major newspapers and television networks stressing the need to control mentally ill people, because of the risks they pose to the public order. In the case of an unmotivated crime in particular, there is not only the suspicion of a mentally ill perpetrator, but all madness is taken as culprit, with the accompanying suggestion of the need to isolate and control it. And the solution in general is to return all power to the psychiatrist and the asylum.[2]

But we wish to follow a different path in this chapter, far removed from those conservative and epic-hagiographical examples. Instead of questioning the directions of the Brazilian Reform Movement, we wish to analyse its devices in a more complex and critical sense. We hope in this way also to contribute to the strengthening of this movement by shedding light on its problems; this is done especially by considering the similarities between some of its practices and neoliberal techniques of government such as entrepreneurship.

Keeping this goal in mind, we will work first with our conception of technoscience, considering psychiatric theories and practices as part of very complex networks, as the actor-network theory suggests. In a more specific way, we will use Latour's concept of 'circulatory system'. Following this, we will explore specific relations present in parts of this 'circulatory system', involving

1 On May 18, 1987 the first congress of the 'Workers Movement of Mental Health' took place in Bauru (a city in São Paulo State). In this conference the following slogan was created: 'For a society without asylums.'

2 In March 2011 there was a shooting in a public school in Rio de Janeiro which ended with the death of 14 students. The murderer left a letter where he justified his crime by saying that he felt a need to purify the victims. Together with expressions of feelings of sadness and perplexity, there immediately appeared in the mass media a great number of analyses featuring mental illness diagnoses of the murderer and ultimate plans for controlling this kind of occurrence.

'public relations', 'alliances', and 'inscription techniques' of these scientific practices. These relations will be analysed by the use of Foucault's genealogical concept introduced at the end of 1970s: governmentality, with a special emphasis on neoliberal management techniques, which build on the freedom of and veridical discourses about the governed. This discussion opens a space for the analysis of the management techniques regarding people's conduct, which are not only present in the old psychiatry, but also in the new reformist process. Within this analysis, we will look for the presence of new risks in the reformist practices and concepts. More specifically, the risks point to the possible similarities between neoliberal management techniques and reformist devices, considering the latter's critical frame. This proximity can be understood by looking at their common effort to overcome the old disciplinary devices through the use of autonomy and self-management. In the conclusion, some alternatives about the concept of freedom will be explored, pointing to a few possibilities not only for psychiatry, but also for our general technoscientific practices.

The circulatory system: An approach toward technoscientific networks

How can we describe these changes in psychiatry? Generally speaking, we have two great alternatives to describe the transformations of a technoscientific device: epistemology and STS. In epistemology, only the internal changes are considered relevant for an evolutionary history of a specific science, which can be explained by the progressive control of our observations (for positivism, both Comtean and from the Vienna School), critical rationality (for Bachelard's and Canguilhem's applied rationalism), or the paradigm crisis (for Thomas Kuhn's paradigmatism). Unlike epistemology, STS analyse the technoscientific transformations in a non-progressive and non-dualistic way, considering inner and external aspects and consequently opening the way to a consideration of socio-political aspects. This non-dualistic position is conducted to a more radical approach in Bruno Latour's actor-network theory, which strives more intensely to escape from any kind of dualism.

Against the idea of progress, and the differentiation between scientific and non-scientific thought, we can find the proposition of the principle of symmetry, suggested by David Bloor and developed by Michel Callon and Latour himself. In this case, this principle is opposed to the concept of fracture and epistemological fissure proposed by applied rationalism, which supposes the existence of a cleft between scientific and common-sense discourse. For Bloor, there is no essential difference between winners and losers in the battle for truth; they should be explained by the same principle, and it is necessary to describe their process of separation. However, Latour and Callon propose a second (general) principle of stronger symmetry, eliminating the modern separation between society and nature and their respective reductions; there is only one socio-nature (Latour 1993). If there is no longer an essential difference between truth and error, science and non-science, then there is also no triumph position of modern knowledge over pre-modern and so-called primitive people.

Not only the ideas of progress and unity, but the concepts of representation and purity are also questioned. For Latour (2004), a technoscientific device arises as an articulation and co-affecting between various actors, with an unexpected production of effects, unlike the representational leap implied in the alleged identity between a sentence and a previous state of affairs that has to be progressively revealed. More specifically, what are the actors that contribute to the creation of a technoscientific device? In spite of the fact that Latour described the specificity of scientific knowledge in several works (Latour 1985, 1992, 1998a, 2006), one text is of special importance: *Science's blood flow: An example from Joliot's scientific intelligence*, which is part of the *Pandora's Hope* collection (Latour 1999a).[3] This article will be taken as representative, since it summarizes a series of contributions in other works within a single model: the *circulatory system*.

And why is scientific work compared to a circulatory system? It is because it makes no sense to ask only about 'the heart of science', but also about the whole, its wide and dense network and capillary system, where articulations are produced. As with our circulatory system it makes no sense for us to ask if in essence it is heart or vein and arteries, with sciences we should not content ourselves only with their conceptual network or social context. This old indictment, maintained by epistemological historians in the venerable debate between internalism and externalism (Latour 2001: 102), will end up imagining scientific knowledge as either produced separately from its social network, like ideas fluctuating in the sky (internalism), or as a mere collective phenomenon, with no understanding of the specificity of sciences (externalism).

In this frame, any technoscientific device will appear as a very strange mix between pure science and social context. The aim of Latour (2001: 103) is to construct articulations between these two supposed universes. Or rather, these two universes are only effects of a purification process. In order to reach this understanding, it is necessary to surpass the a priori division between human and natural actors, only following the translations operated by the scientific networks: 'To discipline men and mobilize things; move things disciplining men: this is the new way to convince, sometimes called scientific search' (Latour 2001: 114).

Real science doesn't work by purifications, by jumps through the gap between our inner mind and the external world. It works through connections and mediations in a rich and constructed world. In other words, it is full of circulatory references. Trying to surpass this 'Berlin wall' between internalists and externalists (and between science and society), and trying to follow the translations and lines of the scientific networks Latour proposes his circulatory system, composed of a series of circuits, such as: 1) *world mobilization*, or the set of mediations able to make different entities circulate through the discourse (instruments, surveys, questionnaires, and expeditions); 2) *autonomization*, or the

3 Hereafter we quote from the Portuguese translation (Latour 2001).

delimitation of a specialist field around a field of study, able to settle or to proceed with a controversy; 3) *alliances*, or recruiting the interest of non-scientific groups such as the military, governmental and industrial sectors; 4) *public representation*, or the set of effects produced around the daily routine of individuals; and 5) *the bonds and knots*, which are related to the conceptual heart that connects all the remaining circuits.

Without the circulation and mobilization of all these circuits, it is not possible to understand the persistence of a scientific work such as Joliot's in his attempt to assemble a neutron bomb. In order to assemble this bomb, not only a network of scientific knowledge is necessary, but also the constitution of laboratories, partnerships with specialists and the government, industry and military interest, besides the public opinion. This is a good example used by Latour (2001) to show his circulatory system model working. But, could this model work to understand a controversial and complex field of knowledge like psychiatry?

'Psychiatric devices' across the circulatory system of sciences: A multiplicity of visions and subjects

First of all, we can say that psychiatry is composed of several circulatory systems as it happens in any scientific discipline. But, differently from other scientific fields the circulatory systems in psychiatry generally do not articulate with each other, working frequently in reciprocal opposition. In this sense, some approaches (as alienisms, organic psychiatry, neuro-biological psychiatry, phenomenological psychiatry, existential psychiatry, psychoanalysis, social psychiatry, etnopsychiatry and so on) have their own circulatory systems, but they do not communicate with each other. As Despret (2004) points out, they work as viewpoints which try to reduce and eliminate any differing perspectives.

Considering these conceptual bonds, we can say that psychiatry is made up of a series of partial conceptual *knots* and *ties* without a bigger knot that binds it all together. What could be the general definition of mental illness (or health) that involves all viewpoints and perspectives? Alienation? Trauma? Degeneration? Neurotransmission disorders? Genetic malfunctioning? Morbid experiences? Unconscious desires? Social conflicts? Each one of these circulatory systems is connected with only a few problems, practical questions, and a network of specific practices, that do not have connections between different viewpoints.

Concerning *autonomization*, among the psychiatrists we have what Canguilhem (1980) calls a more pacific than logic consensus, given the plurality of viewpoints and versions in the field. The autonomy of psychiatry as a discipline and practice arose when a great number of its versions were created without any articulation between them, opening space to a singular inarticulate plurality. Besides this fragmented collective, we can also say that its borders are very porous and open in several directions: neurosciences, social sciences, psychology, pedagogy, social work and law, creating articulated spaces, but also competence disputes.

Regarding *alliances*, we can say that these increase greatly since the definition of psychiatry as a discipline and as practice at the end of the eighteenth century. If in the beginning psychiatry was much more adapted to the needs of the public order (controlling the supposed unreasoning people), soon it was used to explain unmotivated crimes, judge disabled workers, and evaluate mental disability in children (Foucault 2003, 2006a). Nevertheless, in the middle of the twentieth century, a powerful alliance was established with the pharmaceutics industry, which expanded the initial frontiers of psychiatry. If its initial mission was to classify individuals and groups within a normality/abnormality scale, and its operation was restricted mainly to asylums, in the middle of the twentieth century psychiatry spread to all domains of normal life, offering an increasing number of diseases and medicines to the new mentally ill. But, unlike the old patients, these new ones were brought into a more active position, in order to increase their vital and productive impetus. In all respects, these new alliances were anchored in what Foucault (1998) named *biopower,* or later (2006b, 2008) liberal governance.

But these enacted alliances ignore the complexity and plurality of psychiatry, keeping a blind faith in a presumed knowledge about human nature and its morbid states. A faith which is much wider in the field of *public representations*, even when there exists a certain amount of distrust. To take this into account, it is only a matter of becoming aware that some psychiatric guidelines are more widespread as psychoanalysis (or neuropsychiatry in our days); we do not succeed in relating with ourselves or with others without categories such as the unconscious or the Oedipus complex. We cannot understand ourselves without all the psychopathologies offered to us as truths and identities. Homosexuality as an old psychiatric category is a good example of this production of identities. A more recent example of this is the movement of groups such as people who 'hear voices' in their heads, or the autistics which possess some communities on the Internet.

As for *world mobilization*, we can say that the inscription techniques of this knowledge produce (or extract) testimonies not only from objects anymore, but also from subjects, even when these inscription techniques are taken from other sciences, such genetics or biology (Stengers 1989). As we shall see, for Rose (1998) this is the great innovation of psychology: to create inscription techniques for our subjectivity. The problem is that in the psychiatric field the mobilization techniques do not circulate freely in their extension; they only travel in the circulatory system of a determined orientation, where they can be forged. They are not what Latour (1985) calls an *immutable movable*, but a *mutable immovable.* We can say *immovable*, because their concepts only circulate within a certain viewpoint, and *mutable*, because they transform and manufacture the subjects' experience. Although all psychiatric versions possess a strong power to produce subjects, they do not translate between them. For example, psychoanalysts do not translate their clinical findings and devices into a phenomenological and existential language.

A second problem is the style of articulation between researchers and researched entities. Here we can recognize the great strength of psychiatry, because, more than producing free testimonies from subjects, it extorts testimonies (Stengers 1989), manufacturing more than revealing our inner nature. Foucault (1975) also uses a distinction between truth as revelation and production to understand this psychiatric production of identities. More directly he remarks that psychiatry, especially with its interview techniques, operates by extorting the 'true madness' that patients are led to confess. And for Stengers (1989) this is the great innovation produced by Freud and psychoanalysis: it opens the way to the free discourse of subjects without any extortive influence. In any case, this open discourse is, after all, anchored on the truths of the unconscious (Stengers 1989), or on the transferential relation with the analyst as sign of a true relationship (Foucault 1975). As we shall see later, this link between truth and freedom will be the mark of a new kind of patient management: the neoliberal one.

These problems of extortion lead Latour to distinguish between the diverse ways of articulation between human and natural sciences:

> Contrary to non-humans, humans have a great tendency, when faced with scientific authority, to abandon any recalcitrance and to behave like obedient objects offering to the investigators only redundant statements, thus comforting those same investigators in the belief that they have produced robust 'scientific' facts and imitated the great solidity of the natural sciences! (Latour 2004: 217).

If anything unites the different psychiatries, it is their multiple effect in producing what Latour (1998b) calls 'artificial egos'. In Foucauldian terms, psychiatry works as a productive kind of power, generating pathologies and subjectivities. But, at the same time, it obstructs a great number of practices and discourses produced beyond the psychiatric frontiers, such as the religious one.[4] In this sense, this conception agrees with Latour's (1999b) observation that there exists a way to produce 'science' that works by generating subjectivities in a serial way, silencing the voice of an extensive collective: 'science' against the politics of multiplicity. A kind of 'science' that generally extorts the testimony of its subjects, cutting away any possibility of recalcitrance.

To consider this link between psychiatry and politics in a more detailed way, we will examine how psychiatry directs its patients and its public in specific political ways due to its alliances (with the law, city administration, pedagogy, and the pharmaceutical industry). Does psychiatry always maintain restrictive and extortive relations with its subjects? Or can the new reformist tendencies create new political frameworks? Can these new political forms generate a more pluralistic and recalcitrant relation between this knowledge and its technosocial collective? To answer these questions, we will use the concept of governmentality.

4 One interesting exception is the work of Thobie Nathan (1996), who considers in a very symmetric way the versions of psychiatry and other cultural ones, without any effort to reduce the second to the first. He considers all of them devices of influence, which create new versions of subjectivity and madness.

Governmentality and the 'psy tactics'

In two courses given at the end of the 1970s at the Collège de France – *Security, Territory, and Population* (Foucault 2007)[5] and *The Birth of Biopolitics* (Foucault 2008) – Foucault introduces the concept of governmentality in the sense of the strategic control of other people's behaviour. More specifically, this concept refers to 'the set of institutions, procedures, analyses and reflections, the calculations and tactics that allow the exercise of this specific yet very complex form of power, the target of which is population' (Foucault 2006b: 136). Nikolas Rose (1998) applies this concept to the history of the 'psy sciences', which he considers from their origins onward as the exercise of a management technique targeted at the population. The latter is a new object that arises in the eighteenth century due to the emergence of new forms of power. New governance strategies arose during this process: kinds of rationalities among which liberalism is unquestionably the one which most vigorously emerges in the eighteenth century. Here we will present a general overview of this process, with a focus on the comprehension of these new management techniques that were created with the old liberalism, and intensified by neoliberalism.

As Foucault (2006b) points out, a crucial moment in the history of the governmental arts occurs in the sixteenth century, when the so-called governance handbooks appeared. The authors of these handbooks (for instance La Perrière in France, Mayenne in Holland, Hohenthal in Germany) are almost completely unknown to our present political thought. What is the innovation of these handbooks? Until the sixteenth century, the population was not considered an issue or a function of the state. Those governmental books contained instructions on how the state should manage not only the flow of commodities but also that of the individuals. According to Foucault (2006b) these issues gained more prominence due to a process of increased urbanization caused by the migration of rural populations and the decline of death rates. These handbooks appeared at the same time in which a new doctrine and new governmental devices concerning the state arose. This doctrine, the Reason of State, is referred to the strategies devoted to reinforce the strength, harmony, happiness, and peace of the state. Governmental devices were expressed by the police as the assemblage of all resources which concerned these aims. In this context the concept of population became increasingly well defined, and was considered crucial for police strategies, that work by recording and correcting the actions of the individuals. Disciplinary devices were a very distinctive mark of the sixteenth and seventeenth centuries, as they acted in constant vigilance and control of the bodies and actions in institutions such as schools, barracks and hospitals. Afterwards, those devices were incorporated by the state itself in its police form.

In the eighteenth century a new form of governmentality arose, due to physiocracy. As an economic doctrine the latter suggests some limits to the police state, maintaining that market phenomena obey general and natural laws. Thus

5 Hereafter we quote from the Portuguese translation (Foucault 2006b).

the best way to manage the markets is to 'let them go' in the flow of their natural fluctuations. All the efforts to directly manage the markets would result in economic disasters. This is the fundamental change made by liberal governmentality in the disciplinary procedures of the police state. Nevertheless, there occurs a radical change in the liberal strategies. If in the eighteenth century liberalism worked in a critical sense, focusing on the excesses of government by the police state, in the nineteenth and especially in the twentieth century it began to operate with positive management techniques. Foucault (2006b, 2008) and Rose (1998) suggest that the development of liberalism leads to an increased effort in population control on the part of the government, which needs to find devices to guide people's actions in accordance with their natural codes of conduct. Therefore, knowledge of their patterns of action and regularities is necessary in order to guide individuals to become independent and responsible citizens. In other words: to induce self-government. Although in our view this distinction is not made sufficiently clear in Foucault's or Rose's works, we can establish a difference between the first liberalism of the eighteenth century and the more recent forms, concerning governmentality. We can say that the old liberalism (including physiocracy and authors such as Adam Smith) is a critique of the excesses of government by the police state, and that the new liberalism (including the Freiburg and Chicago schools and many other independent strategies) is a positive way of government by the truth, nature and responsibility of the individuals.

According to Rose, it is in the context of the neoliberal government strategies that the 'psy sciences' find the conditions for their development: as a scientifically legitimate field of knowledge, with the free regulation of other people's behaviour. For this author, the history of 'psy knowledges' is linked to the history of government in two ways: 1) through inscription techniques (referred to by Latour) that allowed subjectivities to be managed by governance techniques; 2) through the creation of a variety of policies that intended to guide the behaviour of individuals not only through discipline, but mainly through the freedom and free action of people. The best examples of the first aspect are mental tests, but we can include opinion research, attitude scales and so on. The best example of the second aspect can be taken from the entrepreneurial culture produced during the 1980s (Rose 1998, chapter 7). Here, entrepreneurship is also presented as an 'ethic' procedure, because it combines self-governance with the management of other people. The individual in itself becomes an enterprise, and his or her life is considered as a project the objective of which is to increase the person's human capital. For that it is necessary to work actively on the self and to construct a 'lifestyle', with the ultimate goal of happiness. The 'psy sciences' have a special role in this entrepreneurial project, coordinating the government of the individuals with their self-governance worked through the search for self-realization.

We can also find some entrepreneurial devices in today's psychiatry. For instance, the patients (but also all the 'normal people') are led to manage their supposed mental disease as an enterprise, creating a healthy lifestyle through the management of medicines, symptoms and experiences. A very interesting aspect is the change of the old mental sickness that defined a character (as neurotic, psychotic, maniac and so on) to the new syndromes conceived as disturbances. Here the mental illness is conceived as a non-personal part of the person (and not the entire person), described in the third person and manageable (or self-manageable) by the first person, as a manager or entrepreneur.

In sum, neoliberalism is not exclusively an economic theory or a political criticism of the excesses of government. It is a positive governance technique that has its beginnings in economics, but that little by little is transferred to the domain of the general population, opening up the possibility for the expansion of psychological sciences. And, considered in this framework, it includes some practices proposed not only by the physiocrats and old liberal thinkers (as Adam Smith), but even by the new liberalisms (as the schools of Freiburg and Chicago). All of these liberal governance techniques differ greatly from previous ones, such as the techniques of sovereignty (based on legal devices) and disciplinary techniques (based on the constant recording and control of actions as the police state). They are similar to what Deleuze (1992) called the 'society of control', with an open control with constant management of the 'conduct of others'. According to Rose (1998), the 'psy sciences' had a marked influence on these new liberal governance techniques, especially at the very beginning of the twentieth century, where they acted specifically in the creation of democratic societies by stimulating the citizens to adopt a free and active mode of conduct.

However, the history of psychiatry presents a singular path amidst the 'psy knowledges'. Changes in these governance techniques reveal a general shift from exclusively disciplinary forms (as in the asylums) to those based on sovereignty devices (laws and citizenship rights) present in some psychiatric reform processes. In this context, our aim here is to evaluate specifically the changes in governmentality present in the Italian and Brazilian movements of psychiatric reform. The Italian reform movement is also known as 'democratic psychiatry', and it appeared during the 1960s and 1970s, mostly due to Franco Basaglia's efforts in Trieste and Gorizia (North-eastern Italy). The Brazilian psychiatric reform is greatly inspired by the Italian model, and was developed during the 1980s together with the democratic political reform of the state.

The analysis of governance techniques present in these reform processes will be conducted through the examination of the concepts of citizenship and freedom included in the anti-asylum laws (such as Ministério da Saúde 2002), official documents (decisions regarding these laws, such as Ministério da Saúde 2004a, 2004b), academic texts (of the main authors of the reform process such as Basaglia 1979, 1985, 2005), and practices like the social enterprise (see e.g. Rotelli 1994 and Slavich 1985). This analysis will be conducted by considering the types and styles of governance employed in these discourses and devices. Our hypothesis is that in all these processes there is a coexistence of both the old disciplinary processes (like the asylum practices) and the processes of resistance toward them, the sovereign devices (as human rights), and perhaps the neoliberal forms of management (present in the institutionalization of the 'open doors' devices). After this, we will discuss the concept of freedom that is present in some reformist practices. Our aim is to examine whether these new practices of freedom provide room for recalcitrance and new possibilities of life, or they just lead to a narrow conception of self-management based on an established truth regime.

The birth of psychiatry and human rights

As Foucault points out in his *History of Madness* (2006a), until the end of the eighteenth century asylums, the concept of mental illness and even psychiatry as a discipline did not exist. During the seventeenth and eighteenth centuries there existed only a general health care that considered madness as a 'nervous illness', as well as urban administrative procedures that allowed for the exclusion of 'dangerous individuals'. But this process was carried out according to moral criteria, without any medical reference. These 'dangerous individuals' (prostitutes, beggars, libertines, alchemists, wizards, syphilitics, and even madmen) were put into 'labour houses' in order to be morally corrected.

The birth of psychiatry towards the end of the eighteenth century has a mythical representation: it is that of Philippe Pinel (mythically considered the father of this new knowledge) setting the madmen free from the chains of those old 'labour houses'. This type of image is very similar to others depicting the revolutionary days in France. It could be interpreted as an act of freedom that gives shape to a new humanistic science which is capable of understanding both madness and the human being: psychiatry. This is the traditional and official reading of that image. Nevertheless, as Foucault (2006a) indicates, Pinel's gesture of freedom might also be interpreted as the imprisonment of madness by medical chains: through the idea of an alienated human nature, through a concept of mental illness and through the asylum institution (where the madman's alienated nature has to be revealed and treated).

The birth of psychiatry and alienism represents at first a change in governmental procedures: a shift from a disciplinary regime carried out by the urban administration to a medical and psychiatric power. Nevertheless, this birth is accompanied by another very interesting product of those revolutionary days: the Declaration of Human Rights, a new sovereign device. The creation of this new fraternity of universal rights, based on a universal subject to be governed by his/her own conscience and reason, leads to an exclusion: the alienated and mentally ill are excluded from this humanity. All men are equal due to their universal freedom, reason and conscience, save for the mentally ill. This synchronic birth is not a coincidence: the function of new psychiatry is to restore these universal consciousnesses which have been alienated by madness. The double creation of psychiatry and universal human rights represents the creation of new disciplinary procedures and the exclusion of mental illness from all sovereign devices. According to Birman (1992: 74), this is the structural paradox of the creation of the concept of mental illness: its anthropological singularity entails from the outset the exclusion from this new mode of citizenship as well as from any kind of social contract. Only psychiatry can save the mentally ill, restoring their freedom, conscience, reason, citizenship and open the doors to human rights and sovereignty.

A long jump: The psychiatric reforms

During the nineteenth century, alienism in psychiatry slowly disappeared, opening the space for a variety of organic inclinations, which searched for the bodily mark of mental illness. Nevertheless, this epistemological turn did not change the modes of governance in psychiatry. On the contrary, until the beginning of the twentieth century, asylums only became stronger. There is a change in this process in the period of the great wars, due to a number of reasons: the presence of new psychiatric versions (phenomenological existential, psychoanalytic, social historical), the need of a greater labour force after the wars, the comparison between the asylums and the concentration camps, and even changes in Western governmentality (as hinted above).

A number of small, yet radical metamorphoses arise in the psychiatry of different countries (Great Britain, France, Italy, United States, Canada, etc.), questioning the asylums and even introducing social-communitarian devices. Thus, there appear inner institutional reforms (therapeutic community[6]) and preventionist projects (communitarian psychiatry[7] and sector psychiatry[8]). Nevertheless, a more

---

6 Therapeutic Community refers to the experience initially developed during the 1940s in the Monthfield Hospital (Birmingham, UK) by Main, Bion and Reichman. Basically it points to some institutional reforms, restricted to the asylum, aiming to introduce some democratic and participative practices among the patients (see Amarante 1995). Nevertheless, these changes didn't affect the relation between the asylum and its external space.

7 Preventive or Communitarian Psychiatry began in the 1950s in the United States, and was not restricted only to that country. It was deemed necessary to identify the people more susceptible to mental disorders. This approach was based on the idea that prevention means going to the streets, identifying the habits of people in such a way as to identify vices and 'suspicious' precedents. This was done by means of questionnaires or screenings distributed to the population in order to identify possible candidates for intervention. It should be noted that North-American psychiatry left the milieu of natural sciences in search of a dialogue with sociology and behavioural psychology, as noticed by Gerald Caplan in his 1964 book *Principles of Preventive Psychiatry* (see Amarante 1995).

8 This reform was inspired by some of the French progressive psychiatrists, who demanded the immediate transformation of the French 'insane' asylums, defending the therapeutic vocation of psychiatry. In this perspective, the treatment would have to be accomplished within a social environment, in order to avoid segregation and isolation. The psychiatric institution underwent an intense scrutiny, so that each sector would correspond to a geographic and social area. With this came the possibility for patients to relate with others

radical criticism of psychiatric knowledge and institutions is conducted by the Italian democratic psychiatry and the Brazilian anti-asylum movement. At this point, the political aspects of psychiatry are put into question.

The main premise of these movements was that the mentally ill's political citizenship would be nullified after a long process of social exclusion and psychiatric violence. Their political goal was therefore to undo this long political exclusion, which created a new motto: 'freedom is therapeutic' (Rotelli 1994: 153). The aim was to create new conditions allowing the mentally ill to consider themselves as politically active subjects, governed by initiative and responsibility (Basaglia 2005). In order to achieve this, a number of political changes were proposed in psychiatric institutions in Italy during the 1960s and 1970s, especially through Basaglia's efforts. These efforts were made not only in Gorizia and Trieste, where some concepts such as the social enterprise (see below) were created, but the crucial moment was the so-called 'Basaglia's law' (law n° 180) in 1978, which closed all asylums in Italy and made room for the new, 'open doors' services.

The Brazilian psychiatric reform began in the 1980s, based on this frame of political discussion. This is especially due to the democratization process which occurred after the end of the period of military dictatorship. Whilst in Europe the asylums could be compared to concentration camps, in Brazil they were compared to torture rooms. The effects of the Brazilian reformist movement are very similar to those of the Italian one: a great number of new 'open doors' psychiatric institutions and a new set of laws. On the one side, there is the anti-asylum law decreed in 2001 (the federal law number 10216),[9] and on the other side, there is the creation of a great number of 'open doors' institutions such as CAPS (Psychosocial Attention Centres), therapeutic residences and work cooperatives.[10]

---

of the same culture and social class during their hospitalization. Castel (1981) highlights some obstacles that arose within the reform process. Understood as an articulation between the extra and intra-hospital environment, this psychiatric reform did not contest the asylum as a health device.

9 During the 1990s, a reformist project of the psychiatric system was discussed in the national congress, originating the law number 10216, approved on April 6, 2001. This project dismantles the asylum apparatus based on unwilling confinements. Most of them were paid by the government through the financing of hospital beds in private institutions.

10 All these devices are called substitutive in relation to the central protagonist of the old psychiatric services: the asylum. CAPS offer a great number of therapeutic apparatuses without confinement of the patient, and the therapeutic residences are offered especially to the 'chronic' patients without any alternative of residence (own house, relatives, or friends). Few people live in these residences (which try to avoid an asylum appearance) with the support of mental health professionals.

Examining the new forms of governance through the practices of freedom

In this section we address some of the changes in governmental institutions produced by these new movements in Italy and Brazil. They were born as an assemblage of resistance 'counter-conducts' (Foucault 2006a) against the psychiatric mainstream, especially the asylum and the asymmetric relations between psychiatrists and their patients. Some institutions were established as part of these anti-asylum movements; the 'open doors' institutions replace little by little the old therapies centred on hospitals. However, this institutionalization of the anti-asylum devices is not immediate and without risks; sometimes these 'open doors' systems (as the therapeutic residences) work as a kind of micro-asylum. These micro-asylum forms are the target of constant vigilance by the militants of the anti-asylum movements. At the same time, there is a great displacement in the management practices of the patients now included in a sovereign process, the crowning concept of which is the idea of citizenship. This is not an abstract concept: after the 2001 anti-asylum law any sort of medical compulsory internment needs to be registered in the Prosecuting Council (Ministério da Saúde 2002: Article 8, paragraph 1). The relationship between law and madness that arose in an excluding sense (exclusion of rights and responsibilities) now becomes positive.

In sum, we observe in the reformist process an inversion: the effort of these psychiatric movements is to exclude the disciplinary forms of management (present in asylum devices), for the first time opening a space for the sovereign ones. Nevertheless, our hypothesis is that some forms of neoliberal management could appear at the same time in the institutionalization of these new practices. One could think that this represents a political contradiction, considering the remarkably strong presence of leftist theories in these reform movements. As Foucault (2006b) points out, however, this contradiction is only apparent, considering that leftist thought did not produce any specific kind of governmentality, but only a political theory. Following this, left governments in general use either police devices or neoliberal governance techniques. Furthermore, in order to promote the autonomy of mentally ill patients in the cities, one has to act within the neoliberal frameworks that regulate everyday life. The patient's self-concern about their treatment, the possibility of obtaining a job, his or her responsibilities as a consumer and a citizen, all these elements are required if one is to be considered an autonomous individual. These are the claims of the reformist devices, but also of some neoliberal devices such as the entrepreneurial approach, as was seen previously.

How can we visualize these 'dangerous' frontiers between reformist practices and neoliberal ones? For a more detailed analysis of governmental procedures in reformists practices, we will at first consider the concept (or experience) of freedom proposed by Basaglia (2005) as a therapeutic practice. More than a theoretical discussion, this concept/experience needs to be presented as a communitarian process, where the patients increase their own activity and autonomy through symmetrical relations between them and the health care professionals. In this sense, this work is called 'social enterprise' (Basaglia 2005: 154). The union of all social actors (patients or not) should produce a kind of 'active solidarity within an enterprise logic, an enterprising thing, not to assist people, but to help them comprehend themselves' (Basaglia 2005: 158). In other words, all these social symmetrical relations would open the way to new kinds of enterprising actions. At this point, it is possible to think of the similarities between this social enterprise and the new entrepreneurial subjectivities, as Rose (1998, chapter 7) points out, produced by the new forms of neoliberal management. Here, the same attributes are required: self-management, activity, autonomy, and so on. Emphasizing democratic and symmetrical relations in the devices adopted does not exclude neoliberal forms of management.

Another good example of the ways in which these governance techniques function is the concept of work as a therapeutic device, considered as 'an element of mental health in our society' (Rotelli 1994: 156). It is considered to be therapeutic because it creates 'the status of the citizen-worker for these people' (Rotelli 1994: 156). This new status should produce a real social insertion, by the production of commodities with a marketable quality, a 'true object, not something illusory' (Rotelli 1994: 158). Nevertheless, work practices in the Italian model were not sufficient for social inclusion; consumption and exchange of commodities were also crucial. It was important 'to think that the patient could be more of a consumer than a producer' (Rotelli 1994:157). These practices can get very close to the neoliberal entrepreneurial practices.

Let us take another example, now from Brazil: the Psychosocial Attention Centers (CAPS) as an 'open doors' institution, able to treat mental illness through a network within the community, other services (therapeutic residences) and institutions (school, family etc.). The free movement through this network entails special abilities: self-responsibility and self-governance. Remarkably, these abilities are promoted in CAPS workshops. Their targets are to develop 'a stronger social and family integration, the expression of feelings and problems, the training of bodily abilities, work in productive activities, and the collective exercise of citizenship' in mentally ill patients (Ministério da Saúde 2004a: 20). Mentally ill patients are stimulated by health care workers to have an active role in their own treatment and in the construction of social relations. A practical example of this enhancement of activity is the encouragement provided in order to increase participation in communal assemblies in CAPS. Here, the patient has a real possibility to decide and determine the course of his or her own treatment (Ministério da Saúde 2004a: 17).

All these processes of social inclusion, through work, consumption, political participation, self-investment and so on can become very close to the neoliberal forms of management, especially if these new devices are institutionalized without a constant critical revision, resulting in a mere change of old governmental techniques in the direction of self-management. The asylum remains precisely for those patients who are unable of self-control. Therefore, the question is whether it is possible to create a new kind of citizenship and freedom beyond self-management and other neoliberal governmental practices.

Conclusion

We do not aim to affirm that the main actors of the Italian and Brazilian reform have only included madness in a neoliberal form of management. What we want to highlight is the possible presence of these liberal forms of 'conduct of conduct' in the reformist concepts and approaches, especially in regard to their institutionalization, when the vigour of a concept can give way to the automatism of the daily practice. This is a risk that all reformist movements should keep in mind. But this vigilance demands a kind of special attention. Because, considering the different forms of governmentality, all of them can be assembled together in our common psychiatric apparatus. And it is necessary to observe its minimal effects and fluctuations with careful attention (as Latour's actor-network theory suggests, proposing a myopic sight). As Foucault (2006b) points out, a counter-conduct can become a positive technique of government.

If the goal is to question neoliberal forms of government, it is crucial to think citizenship and freedom in another way. Especially, it is crucial to separate the concept of freedom from its neoliberal meaning, considered as a governmental practice of self-investment, self-constitution and self-management, and also from any fundamental truth as offered for example by psychoanalysis and its concepts of unconscious, desire, psychic conflict and so on. As Larrosa (2000) emphasizes, it is necessary to 'free the freedom' from all those ideas and practices which weaken the meaning of this word. This exercise is crucial for what Foucault calls the 'critical history of the present', opening our actuality to other possibilities that differ greatly from those which are ingrained and crystallized. Freedom here is not a natural attribute of human nature, but an effect of critical practices of resistance and problematization of our more precious truths. It is not utopia, but heterotopia. Freedom is not self-management. On the contrary, 'it is an event, an experimentation, transgression, rupture and creation' (Larrosa 2000: 331). In conclusion, we suggest that the reformist movement can find a vantage point in this critical and heterotopical use of the word freedom: to pay detailed attention to current and specific problems and to bring into question any final and utopian solution for madness. Even the liberating ones.

References

Amarante, P. 1995. *Loucos pela Vida*. Rio de Janeiro: Fiocruz.

Basaglia, F. 1979. *Psiquiatria Alternativa: Contra o Pessimismo da Razão, o Otimismo da Prática*. São Paulo: Editora Brasil Debates.

Basaglia, F. 1985. *A Instituição Negada: Relato de um Hospital Psiquiátrico*. Rio de Janeiro: Graal.

Basaglia, F. 2005. A destruição do hospital psiquiátrico como lugar de institucionalização, in *Escritos Selecionados em Saúde Mental e Reforma Psiquiátrica*, edited by P. Amarante. Rio de Janeiro: Garamond, 23–34.

Birman, J. 1992. A cidadania tresloucada: notas introdutórias sobre a cidadania dos doentes mentais, in *Psiquiatria Sem Hospício: Contribuições ao Estudo da Reforma Psiquiátrica*, edited by B. Bezerra Jr. and P. Amarante. Rio de Janeiro: Relume-Dumará, 71–90.

Canguilhem, G. 1980. What is psychology? *Ideology and Consciousness*, 7, 37–50.

Castel, R. 1981. *Gestion des Risques*. Paris: Editions de Minuit.

Deleuze, G. 1992. Postscript on the Societies of Control. *October*, 59 (Winter), 3–7.

Despret, V. 2004. *Our Emotional Makeup: Ethnopsychology and Selfhood*. New York: Other Press.

Foucault, M. 1975. La casa della follia, in *Crimini di Pace*, edited by F. Basaglia and F. Basaglia-Ongaro. Turin: Einaudi, 151–69.

Foucault, M. 1984. Qu'est-ce que les Lumiéres? *Magazin Littéraire*, 207, 35–9.

Foucault, M. 1998. The History of Sexuality Vol. 1: The Will to Knowledge. London: Penguin.

Foucault, M. 2003. *Abnormal: Lectures at the Collège de France, 1974–1975*. New York: Picador.

Foucault, M. 2006a. *History of Madness*. London: Routledge.

Foucault, M. 2006b. *Seguridad, Territorio, Población*. Buenos Aires: Fondo de Cultura Económica.

Foucault, M. 2007. *Security, Territory, Population: Lectures at the Collège de France*, 1977–1978. New York: Palgrave Macmillan.

Foucault, M. 2008. *The Birth of Biopolitics: Lectures at the Collège de France, 1978–1979*. New York: Palgrave MacMillan.

Larrosa, J. 2000. A libertação da liberdade, in *Retratos de Foucault*, edited by G.C. Branco and V. Portocarrero. Rio de Janeiro: Nau, 328–44.

Latour, B. 1985. Les 'vues' de l' espirit. Une introduction à l'anthropologie des sciences et des techniques. *Culture Technique*, 14, 5–29.

Latour, B. 1992. Give me a laboratory and I will raise the world, in *Science Observed*, edited by K. Knorr-Cetina and M. Mulkay. London: Sage, 141–70.

Latour, B. 1993. *We Have Never Been Modern*. London: Harvester Wheatsheaf.

Latour, B. 1998a. Os filtros da realidade. Separação entre mente e matéria domina reflexões acerca do conhecimento. *Folha de São Paulo, Mais!,* 4 January, 15.

Latour, B. 1998b. Universalidade em pedaços. *Folha de São Paulo, Mais!*, 13 September, 3.

Latour, B. 1999a. Science's blood flow: An example from Joliot's scientific intelligence, in *Pandora's Hope. Essays on the Reality of Science Studies*. Cambridge, MA: Harvard University Press, 80–112.

Latour, B. 1999b. A politics freed from science, in *Pandora's Hope. Essays on the Reality of Science Studies*. Cambridge, MA: Harvard University Press, 236–65.

Latour, B. 2001. O fluxo sangüíneo da ciência: um exemplo da inteligência científica de Joliot, in *A Esperança de Pandora*. Bauru: EDUSC, 97–132.

Latour, B. 2004. How to talk about the body. *Body & Society*, 10(2–3), 205–29.
Latour, B. 2006. Vous avez dit 'scientifique'?, in *Chroniques d'un Amateur des Sciences*. Paris: Presses de Mines, 15–17.
Ministério da Saúde. 2002. Lei n° 10.216, in *Legislação em Saúde Mental*. Brasília: Ministério da Saúde.
Ministério da Saúde. 2004a. *Saúde Mental no SUS: os Centros de Atenção Psicossocial*. Brasília: Ministério da Saúde.
Ministério da Saúde. 2004b. *Residências Terapêuticas: o Que São, Para Que Servem*. Brasília: Ministério da Saúde.
Nathan, T. 1996. Entrevista com Thobie Nathan. *Cadernos de Subjetividade*, 4, 9–19.
Rose, N. 1998. *Inventing Our Selves*. Cambridge: Cambridge University Press.
Rotelli, F. 1994. Superando o manicômio: o circuito psiquiátrico de Trieste, in *Psiquiatria Social e Reforma Psiquiátrica*, edited by P. Amarante. Rio de Janeiro: Fiocruz, 149–69.
Slavich, A. 1985. Mito e realidade da autogestão, in *A Instituição Negada: Relato de um Hospital Psiquiátrico*, edited by F. Basaglia. Rio de Janeiro: Graal, 157–74.
Stengers, I. 1989. *Quem Tem Medo da Ciência?* São Paulo: Siciliano.

PART 3
Neoliberalism, technoscience and the environment

Cler intro

Chapter 7
Neoliberalizing technoscience and environment: EU policy for competitive, sustainable biofuels

Les Levidow, Theo Papaioannou and Kean Birch

Introduction

Since the 1990s the European Union's biofuels policy has espoused several aims: energy security, greenhouse gas (GHG) savings, technology export and rural development. On these various grounds, by 2007 the EU was moving towards statutory targets, i.e. to mandate larger markets for 'renewable energy' including biofuels. However, controversy erupted over harmful environmental and social effects, especially in the global South, where land-use changes were anticipating and supplying a larger EU market for biofuels.

In response to the controversy, policymakers and experts focused blame on 'conventional' biofuels, as if these were a transient yet necessary phase towards sustainable biofuels. When enacted in 2009, statutory targets were accompanied by sustainability criteria. Together these incentives and standards were expected to generate technoscientific innovation towards novel, more competitive and sustainable biofuels by using renewable resources more efficiently. Given these policy assumptions about benign markets fulfilling societal needs via technoscientific innovation, EU biofuel policy can provide a case study for relationships between neoliberalism and technoscience.

This chapter will discuss how EU biofuels policy:

- stimulates new markets for knowledge as well as resources;
- assumes that markets drive beneficent innovation; and thus
- deepens links between markets, technoscience and environment.

The theoretical concept 'neoliberalizing the environment' will be extended to links between technoscience and natural resources.

The chapter is structured as follows: the first section discusses analytical perspectives on neoliberalism and technoscience; second section focuses on EU biofuel policy, examining market drivers and economic imaginaries; third section analyses value chains in biofuels; fourth section looks at expectations for a sustainability technofix; fifth section discusses controversy over EU biofuel targets; and Conclusion summarises the main argument.

Analytical perspectives: Neoliberalism and technoscience

To explore the above issues, the chapter draws on several analytical perspectives linking neoliberalism with technoscience and environment. As elaborated later in this section, such perspectives can be summarised and linked as follows.

Neoliberalism is a complex process that includes ideological, economic and political dimensions, according to Jessop (2002: 435–54). Ideologically, neoliberalism calls for organization of economic, political and social relations through formally free choices of profit-seeking individual actors. Economically, this can happen only through expansion of a liberalized, deregulated market within and across national borders. Politically, neoliberalism implies the roll-back of Keynesian intervention and a commitment to the formal freedom of individual actors in the market.

As a process of institutional change, neoliberalism is driven by an underlying assumption that the 'free market' – conceived as natural, neutral and efficient – *should* be the main means to allocate resources. Hence free markets must be created in order to obtain their benefits (Harvey 2005). As regards the environment, such putative benefits include greater efficiency in natural resource use, better conservation, better evaluation of environmental risks, etc. (Castree 2008).

In neoliberal policy frameworks, 'the market' must be designed (or simply presumed) to optimise resource usage; this is done by stimulating technoscience, regulating its forms or direction, and distributing its societal benefits. Resource problems are attributed to market inefficiency, to be remedied by technological advance, stimulated and constituted by market competition. Beyond simply extracting and processing resources, technological innovation redesigns resources as commodities for appropriation and sale, especially through privatizable knowledge. Such frameworks neoliberalize the environment: in the name of protecting natural resources, they are more readily appropriated and turned into exchange value. Neoliberalism thereby promotes market competition as a driver of technoscientific advance that will enhance resource and thus sustainability (Birch, Levidow and Papaioannou 2010).

Incorporating such neoliberal assumptions, EU policy frameworks promote technoscientific development as a key instrument for international competitiveness. Economic imaginaries represent future markets as benefiting a European economic community, as a basis to mobilize resources and policy support for such markets. This framework also represents 'Europe' as a single unit of market competition, while also promoting integration into global capital. Socio-technical imaginaries present eco-efficiency as reconciling economic competitiveness and environmental sustainability. Each aspect is elaborated in the literature survey that follows.

Market-like rule for extending productivity and plunder

As a central tension within neoliberalism, markets are at once redesigned and naturalized, likewise politically constructed and depoliticized. Neoliberal agendas generally involve the 'mobilization of state power in the contradictory extension and reproduction of market(-like) rule', argue Tickell and Peck (2003: 166). Through 'market-friendly' regulations, neoliberal policies have extended various processes such as privatization, marketization, commodification *etc.*, thereby promoting competitive market relations (Heynen et al. 2007).

Expressing classical economic liberalism, eighteenth-century utilitarians portrayed the market as the natural regulator, complementing the natural liberty of the entrepreneur to trade without interference. Through metaphors of machine and market, this new discourse justified the Enclosures as transforming agricultural land into capital, along with new institutions to police dispossession from common lands. Such ideas undermined the earlier discourse of 'natural law', meaning the natural justice of yeomanry living from their own labour on the land as a common societal resource (Williams 1980: 79).

In such ways, classical economic liberalism espoused 'free markets' as if the state were removing unnatural interference from naturally given market exchange. In practice, freedom to exploit resources depends on various coercive measures by the state in establishing and enforcing property rights. Such arrangements have been naturalized as a 'self-regulating market'. Employers' coercive power takes the form of an apparently free, competitive exchange between buyers and sellers. Yet 'the market has been the outcome of a conscious and often violent intervention on the part of government which imposed the market organization on society for non-economic ends' (Polanyi 2001: 258). By treating potentially everything as commodities to be exploited, 'free-market' liberalism provoked resistance against its destructive effects.

In neoliberalism, by contrast to its classical precedent, market conditions are more explicitly constructed to optimise their beneficent role. Conditions necessary for the success of the neoliberal project include individual property rights that enable and ensure competition between individuals in 'free' markets – i.e., where individual property rights are secured against common or social rights. 'Contrary to classical liberalism, neoliberals have consistently argued that their political program will only triumph if it becomes reconciled to the fact that the conditions for its success must be *constructed*, and will not come about "naturally" in the absence of concerted effort' (Lave, Mirowski and Randalls 2010: 661).

Through a circular reasoning, an omniscient market can be adjusted to redress problems that it has caused. As core principles of neoliberalism,

> The Market is an artifact, but it is an ideal processor of information. Every successful economy is a knowledge economy. [...] The Market (suitably re-engineered and promoted) can always provide solutions to problems seemingly caused by The Market in the first place. (Lave, Mirowski and Randalls 2010: 662–3)

With such expectations for the market, neoliberalism more readily justifies the exploitation of human and natural resources.

The entire history of capital accumulation has depended on dispossessions which subordinate labour to capital and colonize natural resources (Marx 2000). In his concept of primitive accumulation, Marx referred to 'the historical process of divorcing the producer from the means of production'. Entire populations were 'forcibly torn from their means of subsistence', thus expropriating the agricultural producers from the soil (Marx 1976: 875–6).

Marx's concept has been extended to an ongoing process called 'accumulation by dispossession', which primarily relates to the privatization of commons or shared resources (Himley 2008: 443). This trans-historical concept draws present-day analogies with early capitalism:

> A closer look at Marx's description of primitive accumulation reveals a wide range of processes. These include the commodification and privatization of land and the forceful expulsion of peasant populations; conversion of various forms of property rights (common, collective, state, etc.) into exclusive private property rights; suppression of rights to the commons; commodification of labor power and the suppression of alternative (indigenous) forms of production and consumption; colonial, neo-colonial and imperial processes of appropriation of assets (including natural resources); monetization of exchange and taxation (particularly of land); slave trade; and usury, the national debt and ultimately the credit system as radical means of primitive accumulation. [...] All the features which Marx mentions have remained powerfully present within capitalism's historical geography up until now. (Harvey 2003: 145)

More generally, capital accumulation has depended upon 'the endless commodification of human and extra-human nature' (Moore 2010: 391). Although several technical improvements helped to make the steam engine economically viable, its success 'was unthinkable without the vertical frontiers of coal mining and the horizontal frontiers of colonial and white-settler expansion in the long nineteenth century' (Moore 2010: 393). Cheap or nearly free raw materials have been supplied by cheap labour, which remains the ultimate source of surplus value: in expending their own labour, workers produce greater exchange value than the cost of reproducing their labour power.

Here lies a practical contradiction: capital-intensive technological innovation transforms living labour into dead labour (e.g. machinery or materials) and so increases the organic composition of capital, i.e. the ratio of dead labour to living labour. This may be done to reduce capital's dependence on labour and to discipline the labour that remains. By reducing the proportion of living labour, however, innovation tendentially limits surplus value. To overcome this limit, geo-spatial expansion has appropriated more human and natural resources: 'hence the centrality of the commodity frontier in modern world history, enabling the rapid mobilization, at low cost (and maximal coercion), of epoch-making ecological surpluses' (Moore 2010: 393).

Industrialization is popularly associated with technological innovation, as if this were the crucial driver.

> And yet every epoch-making innovation has also marked an audacious revolution in the organization of global space, and not merely in the technics of production. […] The revolutionary achievements were made through plunder as much as through productivity. This dialectic of productivity and plunder works so long as there are spaces that new technical regimes can plunder cheap energy, fertile soil, rich mineral veins. (Moore 2010: 405)

Thus global space must be reorganized for realizing the profitability of technological innovation.

From that critical perspective, technoscientific innovation for more efficient resource usage depends upon and facilitates plunder. Conversely, greater resource usage is driven by greater efficiency, e.g., in extracting and processing raw materials. These causal relations operate in both directions: new opportunities for plunder can drive technoscientific innovation for greater productivity.

Techno-fixes for resource conflicts

The tendency towards plunder is disguised, or even reversed, by a hegemonic neoliberal discourse. According to its promises, greater productive efficiency reduces the need for resources and so helps to conserve them. In a circular logic, market competition becomes an environmental saviour by stimulating gains in efficiency and thus sustainability. Environmentalism has been incorporated into models of market progress: this 'has done far more to smooth the "roll-out" of neoliberalizations than attempts to dismiss or reject environmental concerns outright' (McCarthy and Prudham 2004: 279).

To address sustainability problems, the extension of markets has been linked with a technological fix, whose development 'relies on the coercive powers of competition'. This 'becomes so deeply embedded in entrepreneurial common sense, however, that it becomes a fetish belief that there is a technological fix for each and every problem' (Harvey 2005: 68). Such expectations frame sustainability problems as a technical inefficiency, to be overcome by technoscientific innovation.

Technoscientific innovations have been celebrated for greater efficiency, which have facilitated plunder, especially in the agricultural sector. Multinational corporations have successively colonized 'a multitude of new spaces that could not previously be colonized either because the technology or the legal rights were not available' (Paul and Steinbrecher 2003: 228–9). Since the classical enclosures of the eighteenth century, land access has been obtained by formally withdrawing traditional land rights and/or bypassing them through violence. Such enclosures have been extended by biofuel developments in the global South (Levidow and Paul 2010).

As in earlier historical periods, technoscientific innovation is again promoted as means to alleviate competition for resources and to expand their availability, especially to avoid the conflicts around biofuels. Such conflicts are attributed to inefficiency or mis-management, thus diverting responsibility from market competition and its policy drivers (Franco et al. 2010). By historical analogy, 'new efficiencies are likely to generate further economic incentives for monocultural systems to supply biomass to centralised biorefineries' (Smith 2010: 120; cf. Levidow and Paul 2011).

As indicated by the 'biomass' concept, natural resources are always constructed in particular ways. These reorient biophysical characteristics by devising new knowledge and technologies in order to increase productivity and thereby the accumulation of capital. For a long time, this has meant transforming nature into resources through commodification after extraction; this can be seen as the 'formal subsumption of nature', by analogy to labour exploitation (Boyd, Prudham and Schurman 2001).

Resource use also increasingly involves the 'intensification of biological productivity (i.e., yield, turnover time, metabolism, photosynthetic efficiency)' — or the 'real subsumption of nature'. Nature 'is (re)made to work harder, faster and better'. Yet intensification efforts cannot assume the predictable compliance of nature, whose biophysical characteristics may prove recalcitrant to more efficient use. So there is no way to ensure predictability or control of nature prior to implementing new technologies (Boyd, Prudham and Schurman 2001: 563–4).

Real subsumption of nature exemplifies a wider process of neoliberalizing nature. As politics by other means, this process takes many forms – privatization, marketization, deregulation, reregulation, etc. 'For it involves the privatization and marketization of ever more aspects of biophysical reality, with the state and civil society groups facilitating this and/or regulating only its worst consequences' (Castree 2008: 142–3). Various ecological fixes are devised for the problem of capital accumulation – often in the eco-friendly name of conserving resources, but also in the name of remaking nature.

> These logics show that 'neoliberalism' is, in environmental terms, an apparent paradox: in giving full reign to capital accumulation it seeks to both protect and degrade the biophysical world, while manufacturing new natures in cases where that world is physically fungible. In short, nature's neoliberalisation is about conservation and its two antitheses of destroying existing and creating new biophysical resources. (Castree 2008: 150)

A similar contradiction arises in techno-fixes for environmental problems: such innovations reconceptualize and redesign natural resources for more effective commoditization, while also accounting for such resources in its own market-like image.

As a recent controversial example: biofuels and EU targets have been promoted by new corporate alliances spanning several industrial sectors; they aim to restructure agriculture as a biomass source for diverse industrial products, especially as opportunities for privatizable knowledge. Extra demands on land provoked public and expert controversy about how the necessary biomass could be produced sustainably (Franco et al. 2010, Levidow and Paul 2010). As a means to legitimize biofuel targets, sustainability has been conceptually reduced to carbon accounting. These concepts lie at the nexus of the Low-Carbon Economy, a policy concept fetishizing carbon cycles as the prime indicator of sustainability, e.g. for biofuels.

Although claims for environmental benefits have been questioned, the biofuel controversy has been 'constructed as purely physical debates', especially about carbon accounting.

> The challenge of developing biomass energy systems to reduce carbon emissions is by definition a question of industrial ecology. It requires accounting for the flows of energy and matter (in this case especially carbon) throughout every step of the supply chain, from growing plants to harvesting, processing and transporting them to converting them to useful energy to the disposal of remaining 'wastes' which are ideally reused as new resources. The holistic approach of lifecycle analysis also lends itself to the framing of biomass within the terms of the dominant energy discourse of carbon cycling. (van der Horst and Evans 2010: 180)

Here resources are framed in the image of an accumulation regime, pursuing new commodity frontiers, thus projecting a particular vision of the future.

EU policy imaginaries for technological-market progress

The above technoscientific visions can be analysed as imaginaries – 'representations of how things might or could or should be'. These may be enacted as networks of practices (Fairclough 2010: 266). Imaginaries have many aspects and forms, e.g. socio-technical and/or economic (Jasanoff and Kim 2009, Jessop 2005). Innovation agendas combine socio-technical *and* economic imaginaries as complementary visions of societal progress (Levidow, Birch and Papaioannou 2012). These imaginaries enable and reinforce each other: economic conditions for commoditizing knowledge are presupposed by socio-technical progress and vice-versa. Given this circularity, techno-fixes in neoliberal imaginaries are powerful and difficult to contest; they can become self-fulfilling prophecies. Next such imaginaries will be highlighted in EU policy frameworks.

EU innovation policy has been analysed as socio-technical imaginaries of feasible, desirable futures. According to a report criticizing EU policy, *Taking European Knowledge Society Seriously*, prevalent imaginaries anticipate future technoscientific development as central to societal progress:

> Science and technology in this imaginary are staged unambiguously as the solution to a range of social ills, including the problematic identity of Europe itself. To the extent that S&T are recognised to generate problems, these are cast solely in the form of mistaken technological choices. There is no question about whose definition of society's problems or needs S&T should to address, nor any prior question about who participated in determining what is seen to be a 'worthwhile' (commercially profitable or socially needed?) objective or outcome. Accordingly, the immensely normatively-loaded term 'progress' is black-boxed and its democratic examination is curtailed. (Felt and Wynne 2007: 80)

As the report argues, an 'economics of technoscientific promise' creates beneficent expectations in order to attract resources – financial, human, political, etc. Such expectations conflate technoscientific advance with societal progress.

> It is typically simply assumed that the mere advancement to market of a new product, process or technology is sufficient to demonstrate genuine social 'benefit' – despite well-established understandings of market failures, price externalities and the motivating power of vested interests. (Felt and Wynne 2007: 88)

In such ways, socio-technical imaginaries convey 'unifying narratives of imagined and promised European futures, in order to justify interventions and pre-empt disruptive public responses' (Felt and Wynne 2007: 75). Such futures are deployed to promote specific policy changes and to marginalize dissent.

Related to socio-technical imaginaries are economic ones, e.g. imagining the EU as an economic community which will share the benefits of future markets and competition. One such imaginary is the 'Knowledge-Based Economy' (KBE), a policy agenda which links current markets with future ones. The KBE has been largely shaped by the neoliberal agenda, prioritizing knowledge that can be privatized. Through the KBE policy framework, the state

> is promoting the commodification of knowledge through its formal transformation from a collective resource (intellectual commons) into intellectual property (for example, in the form of patent, copyright and licences) as a basis for generating profits of enterprise and rents for individual economic entities as well as for its own fisco-financial benefit, though there remains scope for counter-hegemonic versions. (Jessop 2005: 159)

Initially focusing on ICT, the KBE agenda has been extended to 'the Knowledge-Based Bio-Economy' (KBBE), emphasizing more efficient ways to use and commoditize renewable raw materials, especially from novel organisms or processes. In the KBBE's dominant imaginary, environmental sustainability is translated into a benign eco-efficient productivity through resources which are renewable, reproducible and therefore sustainable; such resources become 'biomass' for flexible conversion to various industrial products. This imaginary has been elaborated by European Technology Platforms in the agri-food-forestry-biofuels sectors, whose proposals shape research priorities and policy agendas. Together these promote a further privatization and commodification of nature, especially by prioritizing knowledge that can extend intellectual property rights (Birch, Levidow and Papaioannou 2010).

EU economic imaginaries have evolved over time, with tensions between early mercantilist versus later neoliberal forms, whereby the latter have facilitated integration of European capital into global capital as an essential route to competitiveness (van Apeldoorn 2002). Towards this aim, neoliberal policies have extended deregulation, privatization, public-private partnerships, intellectual property rights, etc., thus forcibly creating new market relations. In a European context, the political integration process was conceived as an opportunity to further open up the European region to the globalizing world economy and to accelerate the deregulation and privatization of the European economies, liberating market forces from the fetters of government intervention (van Apeldoorn 2000: 166).

These socio-technical and economic imaginaries are combined in ways shaping societal futures. Neoliberal imperatives – including private property rights, competition, market liberalization, etc., – are promoted as necessary means to facilitate beneficent technoscientific developments. Regardless of whether the latter materialize in commercial form, policy frameworks change institutions, thus shaping societal futures and resource usage along specific lines, while pre-empting alternatives: 'If the model is too simple (as we have argued), the diagnosis and policy measures linked to it will not be productive – but will still shape society' (Felt and Wynne 2007: 19). By drawing on the above concepts, let us examine the case of EU policy framework for promoting and justifying a larger market for biofuels.

EU biofuel policy: Market drivers and techno-economic imaginaries

Since the 1990s EU biofuels policy has featured three main justifications – economic advantage, energy security and greenhouse gas (GHG) savings – for promoting 'sustainable' technologies. The first two issues are discussed here and the third issue in the penultimate section below.

According to many policy documents, biofuels offer more secure energy supplies for Europe, save greenhouse gas (GHG) emissions to address global warming, and promote economic development in the rural places where they are produced (CEC 1997, 2001, EC 2003, Biofrac 2006). The meaning and relative weight of these arguments has changed over time, mainly in response to wider policy agendas and public dissent. From an early concern with energy security, commitments to the Kyoto Protocol became increasingly important, driving efforts to reduce GHG emissions (CEC 1997, 2000).

The renewable energy sector has been expected to promote economic development through the creation of expert knowledge, renewable technologies and thus export opportunities for EU industries as 'world leaders' (CEC 1997: 4). In particular, second-generation biofuels are expected to 'boost innovation and maintain Europe's competitive position in the renewable energy sector' CEC 2007b).

For strategic advice on creating markets for biofuels, the European Commission established the Biofuels Research Advisory Council (Biofrac), representing industrial interests pursuing biofuel innovation. In its future vision, Biofrac claims that: 'By 2030 the European Union covers as much as one fourth of its road transport fuel needs by clean and CO_2-efficient biofuels.' Such eco-efficiency will come from horizontally integrating agriculture with other industrial sectors: 'Integrated biorefineries co-producing chemicals, biofuels and other forms of energy will be in full operation' (Biofrac 2006: 16). Economic development is framed as newly created markets facilitating new sustainable technologies that will yield financial returns.

An imaginary combining economic and environmental sustainability was elaborated in a policy report, *An EU Strategy for Biofuels*. The EU faces an opportunity for global leadership through technoscientific advance:

> In general, the production of biofuels could provide an opportunity to diversify agricultural activity, reduce dependence on fossil fuels (mainly oil) and contribute to economic growth in a sustainable manner. [...] The options, which will be developed, need to be sustainable in economic, environmental and social terms, and bring the European industry to a leading position. (CEC 2006a: 6–7)

Extending that imaginary, EU policy seeks to develop and maintain a competitive advantage for biofuel innovation in global markets. Second-generation biofuels are expected to 'boost innovation and maintain Europe's competitive position in the renewable energy sector' (CEC 2007b: 11). According to the European Commission:

> By actively embracing the global trend towards biofuels and by ensuring their sustainable production, the EU can exploit and export its experience and knowledge, while engaging in research to ensure that we remain in the vanguard of technical developments (CEC 2006a: 5).
>
> In parallel, long-term market-based policy mechanisms could help achieve economies of scale and stimulate investment in 'second generation' technologies which could be more cost effective. (CEC 2006b: 28)

Energy insecurity was always another rationale for promoting biofuels. Use of transport fuel is expected to grow, to become more dependent on imports of fossil fuel, and thus to become less secure: 'there is a particular need for greenhouse gas savings in transport because its annual emissions are expected to grow by 77 million tonnes between 2005 and 2020 – three times as much as any other sector'. Consequently, 'the only practical means' to gain energy security is biofuels, along with efficiency measures in transport, argued the European Commission (CEC 2007b: 2, 7). This imperative was reiterated by DG Tren, the *chef de file* for energy policy:

> The sector is forecast to grow more rapidly than any other up to 2020 and beyond. And the sector is crucial to the functioning of the whole economy. The importance and the vulnerability of the transport sector require that action is taken rapidly to reduce its malign contribution to sustainability and the insecurity of Europe's energy supply. (DG Tren 2009)

Raw material for biofuels was originally meant to come from 'indigenous' sources, especially so that the EU could reduce its dependence on imports and so enhance its energy security (e.g. CEC 1997: 4, CEC 2000, CEC 2006b). But prospective sources were later broadened to developing countries (e.g. CEC 2008). That shift responded to industry projections that half the EU biofuel supply could come from imports by 2030 (Biofrac 2006: 16).

This imaginary naturalizes the increase in EU-wide transport as an objective force that must be accommodated. In practice, the increased demand for transport fuel has been driven by the internal market project and EU subsidy for transport infrastructure (Bowers 1993, Fairlie 1993). In that context, EU biofuel policy evades its own responsibility for unsustainable growth, especially by displacing the problem onto new technoscientific means to satisfy both market demands and environmental sustainability. However, public controversy eventually led to a political compromise in EU policy, linking high targets with sustainability criteria (see penultimate section).

Horizontally integrating value chains, extending commodity frontiers

A biofuels market provides an opportunity to expand industrial agriculture on a global scale, facilitating technological innovation and extending commodity frontiers. This opportunity overlaps with a new policy agenda relaunching the Life Sciences as essential tools for a Knowledge-Based Bio-Economy (KBBE) (DG Research 2005). Although the KBBE has been defined in various ways, the dominant account presents natural resources as renewable biomass amenable to conversion into industrial products via a diversified biorefinery, thus horizontally integrating value chains and linking ecological efficiency with economic development (Becoteps 2011). That is, environmental sustainability becomes dependent upon markets to stimulate technological innovation. These imaginaries inform EU agendas for R&D funding.

Biorefinery: Diversifying production from agro-industrial oil wells

Funded by the European Commission, an international research network has developed research agendas around the biorefinery concept. Since 2006 it has aimed to design new generations of bio-based products derived from plant raw materials that will reach the market place ten to fifteen years later (EPOBIO 2006). Its bio-economy vision changes the role of agriculture, which becomes analogous to oil wells:

> It was noted by DOE [Dept of Energy] and E.U. that both the U.S. and E.U. have a common goal: Agriculture in the 21st century will become the oil wells of the future—providing fuels, chemicals and products for a global community. (BioMat Net 2006)

As a primary means to extract and recompose valuable substances for a biorefinery, 'Biotechnology has the potential greatly to improve the production efficiency and the composition of crops and make feedstocks that better fit industrial needs' (EPOBIO 2006: 8).

The 'diversified biorefinery' takes biotechnology beyond first-generation GM crops to more novel ones. Since the 1980s genetic modification techniques have targeted four major crops—corn, soybeans, oilseed rape (canola) and cotton; the first three have been grown increasingly for animal feed. Now industry can use these crops to produce fuel, while also using the residue to produce animal feed and other industrial co-products. Even without GM crops as feedstock, biorefineries are being designed to diversify inputs and outputs, especially through novel enzymes and processing methods. According to a promotional account, renewable (and therefore sustainable) resources will generate by-products which become inputs for more energy production:

> The integrated diversified biorefinery—an integrated cluster of industries, using a variety of different technologies to produce chemicals, materials, biofuels and power from biomass raw materials agriculture—will be a key element in the future. And although the current renewable feedstocks are typically wood, starch and sugar, in future more complex by-products such as straw and even agricultural residues and households waste could be converted into a wide range of end products, including biofuels. (EuropaBio 2007: 6)

This biorefinery vision poses the dual opportunity and imperative of linking diverse sectors along a value chain providing food, animal feed, energy, novel industrial products, etc. Agricultural raw materials become a universal renewable resource that can be used in more efficient ways, thus contributing to sustainability. According to the predecessor of the Biofuels Technology Platform, in the year 2020:

> Integrated biorefineries co-producing chemicals, biofuels and other forms of energy will be in full operation. The biorefineries will be characterised, at manufacturing scale, by an efficient integration of various steps, from handling

> and processing of biomass, fermentation in bioreactors, chemical processing, and final recovery and purification of the product. (Biofrac 2006: 16)

To reap the benefits, e.g. green energy, society faces an objective imperative of horizontal integration across numerous industrial sectors:

> The production of green energy will also face the exceptional challenge of global industrial restructuring in which the very different value chains of agricultural production and the biorefining industries must be merged with the value chains of the energy providers. (ETP Plants for the Future TP 2007: 33)

This innovation agenda links major agricultural industries – e.g., seed, fertilizer, pesticide, commodities and biotechnology – with the energy sector, including the oil, power, and automotive industries. This imaginary foresees production systems as sustainable by using renewable resources, unlike fossil fuels. To better achieve this sustainable future, industry seeks a flexible horizontal integration by diversifying biomass sources and its potential uses. Their research priorities are promoted by various technology platforms, as invited and funded by the European Commission. In particular the European Biofuels Technology Platform advocates the following research aims:

- Maximization of yield and crop resistance to biotic and abiotic factors (pests, diseases, water scarcity, rising temperatures, etc.).
- Initiate innovative cropping systems to allow efficient, bulk material production for food, feed, fiber and fuel (4F agricultural systems).
- Exploitation of marginal land options (EBTP 2008: SRA-24).

The latter concept 'marginal land' imagines vast land tracts as surplus to food and other needs. In practice, this means land 'marginal' to global value chains, regardless of its importance to local populations. Through the concept of 'marginal land', investors imagine crop cultivation becoming lucrative via novel processes which can more efficiently extract and convert natural resources. According to an NGO critique, however:

> What's seen as marginal land is often land used by marginalized people, by economically weaker sectors of communities, especially women. Much of it is communal land, collectively used by local people who might not have an individual land title, but for whom it is a vital resource for water, feed, food, medicines, fuel and other purposes. (Econexus 2009: 6)

Thus the concept readily disguises new enclosures of arable land and rural populations.

Beyond simply extracting more resources, these are imagined as cornucopian renewable substances, to be enhanced in quantity and market value by redesigning plants through new technologies. The Biofuels Technology Platform develops strategies to optimize valuable products from novel inputs. It requests funds to '[d]evelop new trees and other plant species chosen as energy and/or fiber sources, including plantations connected to biorefineries.' For advanced biofuels, a biorefinery needs: 'Ability to process a wide range of sustainable feedstocks while ensuring energy and carbon efficient process and selectivity towards higher added value products,' e.g., specialty chemicals from novel inputs (EBTP 2008: SRA-23).

Its precursor organization drew an analogy between plant material and crude oil: 'New developments are ongoing for transforming the biomass into a liquid 'biocrude', which can be further refined, used for energy production or sent to a gasifier' (Biofrac 2006: 21). The biocrude metaphor naturalizes the use and redesign of plants as functional substitutes for fossil fuels, and thus for horizontally integrating agriculture with other industries. Here an economic imaginary presents new technologies as market imperatives and opportunities, while a socio-technical imaginary presents markets as drivers of technoscientific progress. The neoliberal emphasis on markets frames the sustainability problem as a technical issue of accessing and optimizing renewable resources, i.e. decomposable biomass. For example, the prospect of (second generation) lignocellulosic fuels illustrates how market opportunities frame technical problems. Lignin in plant cell walls impedes their breakdown, thus limiting the use of the whole plant as biomass for various uses including energy. For agricultural, paper and biofuel feedstock systems, 'lignin is considered to be an undesirable polymer' (EPOBIO 2006: 27) – and so must be redesigned.

Shaping and funding R&D agendas

These imaginaries inform R&D agendas, especially via industry lobbies which favour biomass-to-liquid (BTL) fuel technology for several reasons. BTL offers links with other industries and export markets, as well as a potential basis for multiplying value chains. It also complements the existing transport infrastructure that is locked into liquid fuel technologies. According to the European Biofuels Technology Platform:

> Liquid fuels are the preferred choice for road transport due to their relatively higher energy density and the fact that their use, particularly as blends, is more compatible with existing fuel distribution systems and requires little or no modification to power trains. (EBTP 2008: SRA-1)

Substantial funds have therefore been allocated to R&D agendas focused on novel biofuels under the EU's Framework Programme 7, in both the Energy and Agriculture programmes. Informed by industry's priorities, the EU funded a joint call for proposals on 'Sustainable Biorefineries', initially offering €80m total grants. The overall programme has several aims which include: 'enhancing energy efficiency, including by rationalizing use and storage of energy; addressing the pressing challenges of security of supply and climate change, whilst increasing the competitiveness of Europe's industries' (DG Research/Energy 2006: 4). For the latter aim, second-generation biofuels are expected to 'boost innovation and maintain Europe's competitive position in the renewable energy sector' (CEC 2007a: 11). In these ways, renewable energy is framed as more efficiently linking agriculture with energy for proprietary knowledge in global value chains.

As grounds for greater R&D expenditure, industry has emphasized the closed-loop concept, in that wastes must be continually turned into raw materials for the next stage: 'It will be necessary to optimize closed-loop cycles and biorefinery concepts for the use of wastes and residues in order to develop advanced biomass conversion technology' (EBTP 2010: 7, 16). These novel value chains would depend on significant changes in inputs, processing methods, and outputs.

A successful biorefinery would eventually depend on government subsidies for R&D&D – research and development and demonstration plants. According to speakers at the 2010 stakeholder meeting of the Biofuels Technology Platform, the necessary investment is too costly and commercially risky for the private sector, which therefore requests much more public funds to cover the risks. Testing commercial viability requires an expensive scale-up: 'With an estimated budget of €8 billion over 10 years, 15 to 20 demonstration and/or reference plants could be funded' (EBTP 2010: 26). Indeed, without public funds, such research would not get done in the EU.

As labelled by neoliberalism, this 'market failure' provides an important rationale for state intervention supporting new markets. From a critical perspective, such a policy illustrates how neoliberal strategy socializes risks (e.g. R&D costs) and privatizes benefits (e.g. resulting products). Despite the neoliberal rhetoric of freeing market forces, the public sector has been historically decisive in investing in financially risky new technologies. The effect is to subsidize the industry, socialize the cost and privatize the benefits (Block and Keller 2011).

Along those lines, the Commission had already proposed such a large expenditure programme under the 'sustainable bio-energy Europe initiative', likewise favouring BTL conversion processes within diversified biorefineries (CEC 2009). The public sector faces a potentially enormous investment for a speculative promise – whose successful fulfilment would benefit specific private sectors, aided by indirect subsidy from EU targets and from national measures such as tax incentives. Future value chains depend upon the state funding technological scale-up, as well as creating market conditions to support this technoscientific innovation. Here again are overlapping economic and socio-technical imaginaries, together mobilizing public funds for private gain.

However, BTL has faced much criticism as unsustainable. Biomass conversion into combined heat and power offers greater efficiency and GHG savings than BTL, according to a German report (SRU 2007). Indeed, 'there are better ways to achieve greenhouse gas savings and security of supply enhancements than to produce biofuels. And … there are better uses for biomass in many cases', according to an EC expert report (JRC 2008: 22).

Moreover, according to an NGO, the Commission funds research agendas favouring 'private interests', e.g. agbiotech, GM trees, biofuels and processing techniques for their products. And 'promotion of agrofuel production in Latin America for the European market is likely to lead to further expansion of monocultures, destroying natural habitat and replacing small-scale farming systems' (CEO 2009). This criticism indicates a tension between environmental versus economic sustainability, and thus between different imaginaries. Since 2007 this tension has been highlighted by greater disputes over changes in land use.

Disputing sustainability, imagining a techno-fix

By 2007 biofuel expansion was provoking worldwide controversy over various harmful effects, especially in the global South. Such effects include land grabs, deforestation, community dispossession, more chemical-intensive cultivation methods, etc. Many critics counterposed the term 'agrofuel', highlighting the threat it poses 'because of the intensive, industrial way it is produced, generally as monocultures, often covering thousands of hectares, most often in the global South'. In response to US and EU targets, moreover, 'the rapid development of agrofuel markets is encouraging investment in farming operations by the agrofuel industry, already prospecting developing countries for suitable land for energy crops' (Econexus et al. 2007: 6, 22). For several years, Greens in the European Parliament had been promoting biofuels, but now they more clearly sought to exclude food crops and restrict the biomass source to waste materials that would otherwise have no use (Lipietz 2008).

To undermine the EU's proposal for mandatory targets, critics emphasized the 'carbon debt' that results from directly ploughing up forests or grasslands for newly cultivated land. This generates enormous GHG emissions, equivalent to decades of substituting biofuels for fossil fuels. Beyond direct changes in land use, indirect changes can result from crop substitution across the globe. For example, as the EU's leading biofuel user, Germany draws on domestic or Eastern European sources of oilseed rape; its former food uses generate extra imports of palm oil from more distant sources, especially from Indonesia, where new plantations destroy forests.

In such ways, biofuel production displaces food crops to other places, where a once-off destruction of forest or peatland releases enormous GHG; this 'carbon debt' undermines the GHG savings from biofuels (Fargione et al. 2008, Searchinger 2008). Controversy ensued over the extent of such effects, called 'indirect land-use changes' (ILUC). Land-use change became NGO arguments against the Commission's proposal for a 10% biofuels target by 2020, to be formalized in a new Directive (see next section).

However, these criticisms were turned into an extra argument for pursuing a techno-fix: that is, high targets were necessary to simulate biofuel innovation that would minimize environmental destruction. According to a research network funded by the European Commission:

> At a time when the expansion of first-generation biofuels derived from food crops is causing concern and in some sectors of the public active opposition related to questions of sustainability and competition with food, more emphasis has to be placed on second-generation biofuels (Coombs 2007: 17).

Future novel biofuels are variously described as 'second-generation,' 'next-generation,' 'advanced,' etc. They would use non-food parts of plants, or non-food plants such as grasses, or even algae, as means to avoid extra pressure on fertile arable land. As an extra basis for eco-efficiency, such innovations are expected to use 'marginal land' for growing novel non-food crops and to turn 'bio-waste' into energy. Such resources are seen as 'under-utilized' or 'under-valued', i.e. resources otherwise contributing little to markets.

'Marginal land' would allow novel biofuels to avoid the damage caused by current ones, according to the Trade Commissioner who was also promoting global trade liberalization:

> We have all seen the maps showing the vast tracts of land that would be required to replace petrol to any significant degree. That is why research and development into second generation biofuels that are cleaner, more versatile, and can be used on more marginal land is so important. (Mandelson 2007)

The European Commission's unit which assists developing countries foresees a similar remedy: 'The use of technology must improve production efficiency and social and environmental performance in all stages of the biofuel value chain', as a means to avoid competition for land use (EuropeAid 2009). Numerous policy documents imagine that 'marginal land' is abundantly available for biofuel crops, i.e. that this novel use would make cultivation economically viable but without undermining other land uses (Franco et al. 2010).

Such arguments have provided a rationale for EU biofuel targets – as essential incentives for investment in technological development bringing next-generation biofuels, in turn solving the problems created by the first generation. These expectations for a techno-fix assume or imply that inefficient resource usage causes the sustainability problems of current biofuels. With sufficient market incentives, furthermore, the techno-fix is meant to resolve these issues.

Mandating biofuels targets, stimulating innovation, accounting for carbon

When the global biofuel controversy erupted in 2007, the EU already had voluntary targets for 'biofuels or other renewable fuels for transport' and was discussing proposals to make them mandatory. At its March 2007 meeting the EU Council reiterated support for mandatory targets – subject to biofuel production being sustainable and second-generation biofuels becoming commercially available. The Commission's legislative proposal eventually emphasized 'renewable energy' for transport fuel, thus downplaying biofuels (CEC 2008). As a basis for formalizing and legitimizing mandatory EU targets, warnings about unsustainable biofuels were translated into debates over sustainability criteria. Environmental issues were reduced to 'carbon stock' levels and GHG emissions which could be readily calculated, at least for direct changes in land use; other environmental issues were marginalized.

In the debate over sustainability criteria for 2020 targets, each stakeholder group sought to shape or limit a biofuel market in different ways. The biofuels industry supported high targets with stringent criteria for GHG savings, especially as means to stimulate R&D for future novel biofuels and to guarantee a market for them. Agriculture Ministries generally supported the high targets, especially as an extra support for farmers – but not the stringent criteria for GHG savings, which would exclude most biofuels then being produced in Europe. Environmental NGOs called for lower targets, thus seeking to minimize imports from the global South, as well as stringent criteria to protect environments and livelihoods there.

All the above arguments and pressures converged in a political compromise, the 2009 Renewable Energy Directive (RED), whose mandatory targets aimed to stimulate investment. The preamble emphasized 'opportunities for establishing economic growth through innovation and a sustainable competitive energy policy'; in particular, 'mandatory targets should provide the business community with the long-term stability it needs to make rational, sustainable investments in the renewable energy sector' (EC 2009: 16, 17).

Towards the policy aims, 20% of all energy must come from renewable sources (including biomass, bioliquids and biogas) by the year 2020. Likewise, 10% of total transport fuel must come from renewable energy. Sustainability criteria define which biofuels qualify for the targets: greenhouse gas savings must rise from 35% to 50% in 2016 for existing production and to 60% for new installations in 2017. At the time the RED was enacted, the future 60% criterion was fulfilled only by Brazilian bioethanol.

The RED incorporates wider assumptions about resource conflicts resulting from inefficiency, to be remedied through market-like incentives. GHG savings are double-counted for several categories: co-products which could be used for other energy sources or animal feed; wastes and residues, assuming that they have no other use; and advanced biofuels from non-edible material, assuming that the GHG emissions can be assigned to the edible parts. Together those bonuses were meant to reward and stimulate novel biofuels using non-edible plant material, wastes, etc. and/or generating more co-products. The latter are presumed to reduce pressures on land, as if the market were finite: 'Co-products normally replace animal feed, freeing up land that would otherwise be needed for its production' (CEC 2010: 13). In this imaginary, statutory incentives will stimulate new markets that drive technoscientific innovation towards reducing pressures on natural resources.

Industry expectations for market incentives have led to some disappointment. As it turns out, the bonus in GHG savings favours cheap waste materials as feedstock and advanced biofuels whose production cost is similar to first-generation biofuels. This advantage deters investment in more expensive advanced biofuels, especially those needing novel enzymes. So the RED criteria create a 'market distortion', complains a biofuel representative (Vierhout 2011). Paradoxically, state rules are blamed for creating the wrong type of market and thus distorting a market – which would otherwise not exist. As this complaint reveals, official environmental aims help to justify the fundamental, less explicit aim of subsidizing and creating new markets.

The RED specifies adverse changes in land use which would preclude 'sustainable biofuels'. Producers should avoid 'the conversion of high-carbon-stock land that would prove to be ineligible for producing raw materials for biofuels and bioliquids' (EC 2009: 24). Environmental criteria disqualified any biomass sources from 'highly biodiverse', 'primary forest' and 'continuously forested' areas; the latter were defined by statistical criteria. Compliance will be assessed on the basis of company information, or through voluntary certification schemes or bilateral and multilateral agreements (EC 2009).

Indirect land use change (ILUC) has remained controversial. Some environmental NGOs and Green MEPs proposed to include an extra calculation within the sustainability criteria. Instead the issue was deferred: under the RED, by December 2010 the Commission must report on ways to calculate ILUC and to minimize its impact (EC 2009, see end of this section for outcome in CEC 2010).

Also at issue was social sustainability. A Parliamentary committee had proposed that sustainability criteria should include social aspects, e.g. land rights of local communities and fair remuneration of workers. But these criteria were ultimately excluded, partly on grounds that they would contravene WTO rules on trade barriers (EP Envi 2008, Biofuelwatch et al. 2008). 'These directives do not include mandatory social criteria (labour conditions, land tenure, etc.), nor food security criteria, because of the difficulty to verify the link between individual biofuel consignments and the respect of these particular criteria', according to a Commission development agency (EuropeAid 2009: 2). Indeed, complex trade flows leave no one responsible for harmful consequences.

Novel future biofuels were meant contribute significantly to the 10% target, yet these expectations were soon contradicted by the aggregate National Renewable Energy Action Plans (NREAPs). Their implications were analysed in a NGOs' joint report which was ominously entitled, *Driving to Destruction*. According to its analysis, conventional biofuels would contribute up to 92% of total predicted biofuel use, representing 8.8% of the total energy in transport by 2020. Moreover, '72% of this demand is anticipated to be met through the use of biodiesel and 28% from bioethanol' (Bowyer 2010: 2) – significant because biodiesel causes relatively greater harm than bioethanol via ILUC effects.

On that basis, the NGOs' report questioned the 10% target. They warned that emissions from ILUC will be '80.5 to 166.5% worse than would be delivered from continued reliance on fossil fuels in the transport sector', especially as the EU uses more biodiesel. 'Moreover, it also raises urgent questions about the appropriateness of projected levels of conventional biofuel use by Member States in 2020' (Bowyer 2010: 2). Several NGOs warned that many decades or even centuries may be needed to repay the 'carbon debt'. Meanwhile this debt is concealed by 'carbon laundering' through statutory criteria which account only for direct changes in land use (Birdlife International 2010).

In response to the ILUC controversy, the Commission held a public consultation (DG Energy 2010). NGOs argued that ILUC effects warrant more stringent sustainability criteria, in order to justify the 10% target as environmentally beneficial. Industry argued that available ILUC models suffer from methodological weaknesses, thus providing no basis for extra regulatory measures. And industry warned that such measures would jeopardize biofuel investment (ILUC 2010).

Afterwards the Commission left open its future policy options. Its report reiterated the EU's beneficent expectations for future technological innovation, which depends on incentives for private-sector investment:

> Biofuels are important because they help tackle two of the most fundamental challenges in energy policy with regards to transport: the overwhelming dependency of the transport sector for oil and the need to decarbonise transport. Supporting biofuels offers other opportunities too. They can contribute to employment in rural areas, both in the EU and in developing countries and they offer scope for technological development, for example in second-generation biofuels. [...] In this context the stable and predictable investment climate created by the Renewable Energy Directive [...] needs to be preserved. (CEC 2010: 2, 14)

In this circular reasoning, echoing industry stances, EU policy must maintain profit-seeking incentives to generate future innovation alleviating the harm caused by current markets. Industry expectations to profit from future innovation, combining socio-technical and economic imaginaries, supersede governments' acknowledgement that advanced biofuels will make little contribution by 2020.

Conclusion: Neoliberalizing technoscience and environment

Let us return to the original focus — how EU biofuels policy:

- stimulates new markets for knowledge as well as resources;
- conceptualizes and designs markets as a driver of beneficent innovation; and thus
- deepens links between technoscience and neoliberalism.

EU biofuels policy promotes a vision of a feasible, desirable future Europe constituted by economic and socio-technical imaginaries. Namely, market-driven innovation will generate 'competitive, sustainable biofuels' within a wider Knowledge-Based Bio-Economy (KBBE). This will achieve a benign eco-efficient productivity using resources which are renewable, reproducible and therefore sustainable: such resources (especially non-food biomass) will replace fossil fuels. Future biofuel production will efficiently use renewable resources to enhance energy security, economic competitiveness, technology export and GHG savings – aims which already drive innovation, according to the imaginary.

As a key imperative for biofuels, EU policy foresees greater future demand for oil imports and thus energy insecurity, as if this were an objective external force. In practice, such pressures result from long-standing market-based policies (e.g. transport infrastructure, internal market, trade liberalization, etc.) throwing people and products into greater competition with each other. In this neoliberal context, EU biofuels policy naturalizes energy insecurity, which is attributed to external pressures such as oil dependence. Likewise greater pressures on land and natural resources are naturalized; they are attributed to global market demand, as if this were external to the production system for food, feed and energy. Within those problem-definitions, biofuels are promoted as a multi-purpose remedy, which can be characterized as a technological fix: future efficiency improvements will sustainably expand biofuel production, while minimizing demands on natural resources. This agenda promotes new markets for natural resources, while discursively naturalizing those markets as rooted in biological characteristics and objective imperatives.

That relationship between neoliberalism and technoscience seems to shape the EU policy framework. To construct new markets for biofuels, EU policy has two complementary means: biofuel targets and R&D subsidy, which can be summarized as follows. As a statutory target, 10% of transport fuels must come from renewable energy by 2020 under the 2009 RED. The target mandates a significant market which otherwise would hardly exist, in the name of environmental benefits. These markets are regulated by sustainability criteria, incorporating only those issues which can be calculably reduced to carbon accounting. Through profit-seeking incentives and standards, the RED aims to stimulate investment and innovation to fulfil the target, eventually through more sustainable biofuels to comply with stricter standards for installations built after 2017. In practice, the draft RED was already stimulating land grabs anticipating opportunities to supply new EU markets.

In parallel, industry promotes research agendas for horizontally integrating agriculture with other industries, including energy production, especially through an integrated biorefinery. Its agenda favours biomass-to-liquid technologies, which offer various prospects for privatizing knowledge, as a central feature of the broader Knowledge-Based Economy agenda (cf. Jessop 2005). R&D agendas redesign nature for real subsumption to capital accumulation (cf. Boyd, Prudham and Schurman 2001), e.g. by changing cell-wall composition for easier breakdown or entire plants for high-value substances. At the same time, industry requests enormous subsidy on grounds that the financial risks of early-stage development are too great for private investors. The EU's Framework Programme 7 has incorporated these neoliberal imaginaries into research priorities, e.g. for 'sustainable biorefineries'.

As an economic imaginary around biorefineries, 'value chains' help to mobilize political, financial and organizational investment for biofuels R&D. New cross-sectoral industry coalitions imagine an economic community gaining together from future technological development. Given various competing interests, intra-EU rivalry for global capital integration is represented as 'European competitiveness', as if Europe were a unitary interest (cf. Rosamond 2002).

R&D anticipates a diversified biorefinery integrating agriculture with other industrial sectors. Investors seek an advantageous position in future global value chains from agriculture, seen as new 'oil wells' whose biomass can be 'cracked' and recomposed into more valuable components. Future biofuels are also promoted as an opportunity for European technology export, e.g. agri-inputs and biomass processing techniques which can be patented. In these ways, the 'value chains' concept combines economic imaginaries with socio-technical imaginaries to stimulate new investments.

These imaginaries extend a cornucopian vision of resources for lucrative biomass, especially via its redesign, diversification and recomposition for multiple uses. Vast areas are imagined as 'marginal land', unnecessary for food production and so benignly available for agro-industrial systems. This concept helps to justify EU targets which stimulate changes in land use, despite causing environmental destruction and dispossessing local populations. These changes exemplify capital accumulation by dispossession (Harvey 2003), whereby investors gain access to cheap human and natural resources at new commodity frontiers (Moore 2010). EU targets have remained contentious, for several reasons. The market drivers of harm have been highlighted by many NGO critics, thus politicizing EU targets. These stimulate harmful land-use changes, especially in in the global South. Despite expectations and incentives for novel biofuels, nearly the entire 10% target by the year 2020 will come from conventional biofuels, thus stimulating direct and indirect changes in land use. These create an enormous 'carbon debt', which plausibly undermines GHG savings and renders most biofuels 'unsustainable', thus contradicting the official environmental rationale. Indirect land-use change (ILUC) has been deferred from any statutory rules, which fall within the standard disciplines of trade liberalization, thus imposing a great burden of evidence to justify any extra 'discriminatory' criteria.

Meanwhile the EU biofuel controversy has been channelled into disputes over carbon accounting and its methodological difficulties in predicting environmental effects. Those difficulties are turned into policy deference to neoliberal imperatives, especially an 'investment climate' for market competition, as the implicit knowledge-base for technological solutions (cf. Lave, Mirowski and Randalls 2010), thus depoliticizing EU targets. In such ways, environmentalism has been incorporated into models of market progress, as a more effective neoliberal strategy than simply disregarding environmental issues (cf. McCarthy and Prudham 2004).

In all those ways, EU biofuel policy illustrates the joint neoliberalization of technoscience and the environment. Through circular reasoning, incentives for profit-seeking investment must be maintained in order eventually to achieve the 10% target, in innovative ways avoiding the harm caused by current biofuels. Such incentives are meant to generate 'competitive, sustainable biofuels' – by stimulating technoscience, regulating its forms or direction, distributing its societal benefits and optimizing resource usage. In practice, EU targets stimulate capital accumulation by socio-economic dispossession and more GHG emissions; more 'efficient' innovations may provide even greater incentives. Official environmental aims help to justify the less explicit aim of subsidizing and creating new markets.

This agenda is depoliticized as a benign, omniscient market – justified and guided by sustainability criteria – as the most efficient mechanism to achieve environmental goals. Any political accountability is reduced to carbon accounting, in turn relegated to specialists, while marginalizing other knowledges. Thus the EU policy framework facilitates plunder and commoditization of natural resources – in the name of conserving them, perhaps like neoliberalization processes in general.

Acknowledgements

Research leading to these results has received funding from the European Community's Seventh Framework Programme under grant agreement n° 217647. Entitled 'Co-operative Research on Environmental Problems in Europe' (CREPE, www.crepeweb.net). Helpful editorial comments were provided by David Hess, Luigi Pellizzoni and Marja Ylönen.

References

Becoteps. 2011. *The European Bioeconomy in 2030: Delivering Sustainable Growth by addressing the Grand Societal Challenges*. European Commission, 7th Framework Programme, BECOTEPS – Bio-Economy Technology Platforms Available at: http://www.epsoweb.org/file/560. [accessed: 18 December 2011].

Biofrac. 2006. *Biofuels in the European Union: A Vision for 2030 and Beyond.* Final draft report of the Biofuels Research Advisory Council.

BioMat Net. 2006. 1st International Biorefinery Workshop. [website defunct].

Birch, K., Levidow, L. and Papaioannou, T. 2010. Sustainable capital? The neoliberalization of nature and knowledge in the European 'Knowledge-Based Bio-Economy'. *Sustainability* [Online], 2(9), 2898–2918. Available at: http://www.mdpi.com/2071-1050/2/9/2898/ [accessed: 18 December 2011].

Birdlife International. 2010. *Bioenergy: A Carbon Accounting Time Bomb* [Online]. Birdlife International, with European Environment Bureau and Transport & Environment. Available at: http://www.birdlife.org/eu/pdfs/carbon_bomb_21_06_2010.pdf [accessed: 18 December 2011].

Block, F. and Keller, M.R. 2011. *State of Innovation*. London: Paradigm.

Bowers, C. 1993. Europe's motorways. *The Ecologist*, 23(4), 125–30. Available at: http://exacteditions.theecologist.org/exact/browse/307/308/5346/2/1/0/ [accessed: 18 December 2011].

Bowyer, C. 2010. *Anticipated Indirect Land Use Change Associated with Expanded Use of Biofuels and Bioliquids in the EU: An Analysis of the National Renewable Energy Action Plans*. Brussels: Institute for European Environment Policy.

Boyd, W., Prudham, S. and Schurman, R. 2001. Industrial dynamics and the problem of nature. *Society and Natural Resources*, 14, 555–70.

Castree, N. 2008. Neoliberalising nature: The logics of deregulation and reregulation. *Environment and Planning A* 40, 131–52.

CEC. 1997. *Energy for the Future: Renewable Sources of Energy*. Communication from the Commission. White Paper for a Community Strategy and Action Plan. COM(97) 599 final. Brussels: Commission of the European Communities.

CEC. 2000. *Green Paper: Towards a European Strategy for the Security of Energy Supply*. Available at: http://ec.europa.eu/energy/green-paper-energy-supply/doc/green_paper_energy_supply_en.pdf [accessed: 18 December 2001].

CEC. 2001. *Communication from the Commission on a proposal for a directive of the European Parliament and the Commission on the promotion of the use of biofuels for transport*. COM(2001) 547. Brussels: Commission of the European Communities.

CEC. 2006a. *An EU Strategy for Biofuels*. Communication from the Commission. COM(2006) 34 final. Brussels: Commission of the European Communities.

CEC. 2006b. *An EU Strategy for Biofuels – Impact Assessment. Commission Staff Working Document: Annex to the Communication from the Commission*. SEC(2006) 142.

CEC. 2007a. *An Energy Policy for Europe*. Communication from the Commission to the European Council and the European Parliament. COM(2007) 1 final. Brussels: Commission of the European Communities.

CEC. 2007b. *Biofuels Progress Report: Report on the Progress Made in the Use of Biofuels and Other Renewable Fuels in the Member States of the European Union*. SEC(2006) 1721. Brussels: Commission of the European Communities.

CEC. 2008. *Proposal for a Directive of the European Parliament and of the Council on the Promotion of the Use of Energy from Renewable Sources*. January. Brussels: Commission of the European Communities.

CEC. 2009. *A Technology Roadmap for the Communication on Investing in Development of Low-Carbon Technologies (SET-Plan)*. Commission Staff Working Document. Brussels: Commission of the European Communities.

SEC(2009) 1295. Available at: http://ec.europa.eu/energy/technology/set_plan/set_plan_en.htm. [accessed: 18 December 2011].

CEC. 2010. *Report from the Commission on Indirect Land-Use Change Related to Biofuels and Bioliquids*. COM(2010) 811 final. Brussels: Commission of the European Communities. Available at: http://ec.europa.eu/energy/renewables/biofuels/land_use_change_en.htm [accessed: 18 December 2011].

CEO. 2009. *Agrofuels and the EU Research Budget: Public Funding for Private Interests*. Brussels: Corporate Europe Observatory. Available at: http://www.corporateeurope.org/agrofuels/content/2009/05/agrofuels-and-eu-research-budget [accessed: 18 December 2011].

Biofuelwatch et al. 2008. *Sustainability Criteria and Certification of Biomass – Greenwashing Destruction in Pursuit of Profit*. Joint press release. Biofuelwatch, Corporate Europe Observatory, Econexus, Grupo Reflexion Rural, NOAH (Friends of the Earth Denmark). Available at: http://www.biofuelwatch.org.uk/files/jointngopressrelease180308.pdf [accessed: 18 December 2011].

Coombs, J. 2007. *Building the European Knowledge-Based Economy: The Impact of 'Non-Food' Research (1988 to 2008)*. Speen: CPL Scientific Publishing.

DG Energy. 2010. *Indirect Land Use Change Impacts of Biofuels – Consultation, 2 August*. Brussels: European Commission, Directorate-General for Energy. Available at: http://ec.europa.eu/energy/renewables/consultations/2010_10_31_iluc_and_biofuels_en.htm [accessed: 18 December 2001].

DG Research. 2005. *New Perspectives on the Knowledge-Based Bio-Economy: Conference Report*. Brussels: European Commission, Directorate-General for Research & Innovation.

DG Research/Energy. 2006. *FP7 Theme 5: Energy, 2007 Work Programme*. Brussels: European Commission, Directorates-General for Research & Innovation and for Energy.

DG Tren. 2009. *Biofuels and Other Renewable Energy in the Transport Sector*. Brussels: European Commission, Directorate-General for Transport and Energy [later re-named Energy]. Available at: http://ec.europa.eu/energy/renewables/biofuels/biofuels_en.htm [accessed: 18 December 2011].

EBTP. 2008. *Strategic Research Agenda & Strategy Deployment Document*. European Biofuels Technology Platform. Speen: CPL Press.

EBTP. 2010. *Strategic Research Agenda 2010 Update: Innovation Driving Sustainable Biofuels*. European Biofuels Technology Platform. Available at: http://www.biofuelstp.eu/srasdd/SRA_2010_update_web.pdf [accessed: 5 June 2012].

EC. 2003. Directive 2003/30/EC of the European Parliament and of the Council of 8 May 2003 on the promotion of the use of biofuels or other renewable fuels for transport. *Official Journal of the European Union*, L 123, 17 May, 42–6.

EC. 2009. Directive 2009/28/EC of the European Parliament and of the Council of 23 April 2009 on the promotion of the use of energy from renewable sources and amending and subsequently repealing Directives 2001/77/EC and 2003/30/EC Renewable Energy Directive. *Official Journal of the European Union*, L 140, 5 June, 16–62.

Econexus et al. 2007. *Agrofuels: Towards a Reality Check in Nine Key Areas*. Available at: http://www.econexus.info/sites/econexus/files/Agrofuels.pdf [accessed: 18 December 2011].

Econexus et al. 2009. *Agriculture and Climate Change: Real Problems, False Solutions*. Report published for the Conference of the Parties (COP15) of the United Nations Framework Convention on Climate Change in Copenhagen, December. Available at: http://www.econexus.info/sites/econexus/files/Agriculture_climate_change_copenhagen_2009.pdf [accessed: 18 December 2011].

EP Envi. 2008. *Sustainability Criteria for Biofuels: Workshop Consolidated Texts*. Brussels, 4 March, PE 404.887. Brussels: European Parliament, Committee on the Environment, Public Health and Food Safety.

EPOBIO. 2006. *Products From Plants – The Biorefinery Future. Outputs from the EPOBIO Workshop*. Wageningen, 22–24 May 2006, EPOBIO project (Realising the Economic Potential of Sustainable Resources – Bioproducts from Non-food Crops), jointly funded by FP6 and US Dept of Agriculture. Available at: http://www.epobio.net/workshop0605.htm [accessed: 18 December 2011].

ETP Plants for the Future. 2007. *European Technology Platform Plants for the Future: Strategic Research Agenda 2025. Part II*. Brussels: EPSO.

EuropaBio. 2007. *Biofuels in Europe: EuropaBio Position and Specific Recommendations*. Brussels: European Association for Bioindustries. Available at: http://www.europabio.org/sites/default/files/position/biofuels_in_europe.pdf [accessed: 18 December 2011].

EuropeAid. 2009. *Position on Biofuels for the ACP-EU Energy Facility*. European Commission, Directorate-General for Development and Cooperation. Available at: http://ec.europa.eu/europeaid/where/acp/regional-cooperation/energy/documents/biofuels_position_paper_en.pdf [accessed: 18 December 2011].

Fairclough, N. 2010. *Critical Discourse Analysis: The Critical Study of Language*. 2nd Edition. London: Pearson.

Fairlie, S. 1993. The infrastructure lobby. *The Ecologist*, 23(4), 123–4.

Fargione, J., Hill, J., Tilman, D., Polasky, S. and Hawthorne, P. 2008. Land clearing and the biofuel carbon debt. *Science*, 29 February, 1235–8.

Felt, U. and Wynne, B. 2007. (Eds.) *Taking European Knowledge Society Seriously*. Report for the European Commission. Luxembourg: Office for Official Publications of the European Communities. Available at: http://ec.europa.eu/research/science-society/document_library/pdf_06/european-knowledge-society_en.pdf [accessed: 5 June 2012].

Franco, J., Levidow, L., Fig, D., Goldfarb, L., Hönicke, M. and Mendonça, M.L. 2010. Assumptions in the European Union biofuels policy: Frictions with experiences in Germany, Brazil and Mozambique. *Journal of Peasant Studies*, 37(4) (Special Issue: Biofuels, Land and Agrarian Change), 661–98.

Harvey, D. 2003. *The New Imperialism*. Oxford: Oxford University Press.

Harvey, D. 2005. *A Brief History of Neoliberalism*. Oxford: Oxford University Press.

Heynen, N., McCarthy, J., Prudham, S., Robbins, P. 2007. (Eds.) *Neoliberal Environments: False Promises and Unnatural Consequences*. New York: Routledge.

Himley, M. 2008. Geographies of environmental governance: The nexus of nature and neoliberalism. *Geography Compass*, 2(2), 433–51.

ILUC. 2010. *Submissions from UK, NL, FoEE, EBB, etc. in response to DG Energy (2010)*. Available at: http://ec.europa.eu/energy/renewables/consultations/2010_10_31_iluc_and_biofuels_en.htm [accessed: 18 December 2011].

Jasanoff, S. and Kim, S.-H. 2009. Containing the atom: Sociotechnical imaginaries and nuclear power in the United States and South Korea. *Minerva*, 47(2), 119–46.

Jessop, B. 2002. Liberalism, neoliberalism, and urban governance: A state-theoretical perspective. *Antipode*, 34(3), 452–72.

Jessop, B. 2005. Cultural political economy, the knowledge-based economy, and the state, in *The Technological Economy*, edited by A. Barry and D. Slater. New York: Routledge, 144–66.

JRC. 2008. *Biofuels in the European Context: Facts and Uncertainties*. Ispra: European Commission, Joint Research Centre, JRC 43285.

Lave, R., Mirowski, P and Randalls, S. 2010. Introduction: STS and neoliberal science. *Social Studies of Science*, 40(5), 659–675.

Levidow, L. and Paul, H. 2010. Global agrofuel crops as contested sustainability, Part I: Sustaining what development? *Capitalism Nature Socialism*, 21(2), 64–86.

Levidow, L. and Paul, H. 2011. Global agrofuel crops as contested sustainability, Part II: Eco-efficient techno-fixes? *Capitalism Nature Socialism*, 22(2), 27–51.

Levidow, L., Birch, K., Papaioannou, T. (2012). EU agri-innovation policy: Two contending visions of the Knowledge-Based Bio-Economy. *Critical Policy Studies,* 6(1), 40–65.

Lipietz, A. 2008. Food or fuel? *Research Review*, May 2008 [European Parliament magazine]. Available at: www.e-pages.dk/dods/12/15, http://lipietz.net/spip.php?article2242 [accessed: 18 December 2011].

McCarthy, J. and Prudham, S. 2004. Neoliberal nature and the nature of neoliberalism. *Geoforum*, 35(3), 269–393.

Mandelson, P. 2007. *The Biofuel Challenge*. Speech at Biofuels conference, 5 July. Brussels: European Commission, Directorate-General Trade.

Marx, K. 1976. *Capital*, Volume 1. London: Penguin.

Marx, K. 2000. Theories of surplus value, in *Karl Marx Selected Writings*, edited by D. McLellan. Oxford: Oxford University Press.

Moore, J. 2010. The end of the road? Agricultural revolutions on the capitalist world-ecology, 1450–2010. *Journal of Agricultural Change*, 10(3), 389–413.

Paul, H. and Steinbrecher, R. 2003. *Hungry Corporations: Transnational Biotech Companies Colonize the Food Chain*. London: Zed.

Polanyi, K. 2001. *The Great Transformation*. Boston: Beacon Press (first edition 1944).

Rosamond, B. 2002. Imagining the European economy: 'Competitiveness' and the social construction of Europe' as an economic space. *New Political Economy*, 7(2), 157–77.

Searchinger, T. 2008. *Summaries of Analyses in 2008 of Biofuels Policies by International and European Technical Agencies*. GMF-The German Marshall Fund of the United States, Economic Policy Program. Available at: http://www.eepa.co.za/docs/GMF%20brief%20summary%20on%20european%20analyses%20for%20biofuels.pdf [accessed: 18 December 2011].

Smith, J. 2010. *Biofuels and the Globalisation of Risk*. London: Zed.

SRU. 2007. *Climate Change Mitigation by Biomass*. Special Report. Sachverständigenrat für Umweltfragen (SRU), Advisory Council on the Environment. Available at: http://eeac.hscglab.nl/files/D-SRU_ClimateChange Biomass_Jul07.pdf.

Tickell, A. and Peck, J. 2003. Making global rules: Globalization or neoliberalization?, in *Remaking the Global Economy*, edited by J. Peck and H. Yeung. London: Sage, 163–81.

van Apeldoorn, B. 2000. Transnational class agency and European governance: The case of the European Roundtable of Industrialists. *New Political Economy*, 5(2), 157–81.

van Apeldoorn, B. 2002. *Transnational Capitalism and the Struggle over European Integration*. New York: Routledge.

van der Horst, D. and Evans, J. 2010. Carbon claims and energy landscapes: Exploring the political ecology of biomass. *Landscape Research*, 35(2), 173–93.

Vierhout, R. 2011. *Policy Framework, Including NREAPs, Targets for Commercial and Advanced Biofuels*. Talk at 4th Stakeholder Conference of European Biofuels Technology Platform (EBTP), Brussels 14–15 September.

Williams, R. 1980. Ideas of nature, in *Problems in Materialism and Culture*. London: Verso, 67–85.

Chapter 8

Configuring homo carbonomicus: Carbon markets, calculative techniques, and the green neoliberal

Anders Blok

Introduction: Carbon economies of hopes and failures

When thinking about the social life of techno-scientific objects, carbon is a good candidate for an object that was 'born neoliberal'. Since the middle of the 1990s, closely intertwined with the rise of widespread concerns over climatic risks in science, politics and mass media, massive investments have gone into forging an entirely new world economy centred around carbon as a tradable commodity (Goodman and Boyd 2011). Across scales and contexts, the resulting carbon markets are commonly depicted, in social science discourse, as expressing and heralding a global technocratic (dis-)order of neoliberal environmental governance (e.g. Lohmann 2010). Such critical assessments resonate also in wider publics: leading up to the 2010 Toronto G20 meeting, for instance, one Canadian community of journalists and activists ran an on-line poll, asking their readers if 'carbon trading [is] just neoliberalism under a green veil?'. So far, just about half the people in this micro-public agree, calling for the elimination of carbon trading.[1]

While anecdotal, the poll serves to index how carbon markets have emerged, in the disconcerting context of global warming, as core sites of the contentious entanglement of new techno-scientific knowledge, neoliberal market-based policies, and public concerns with environmental risks (Blok 2011). Far from any (neo-)liberal ideological fantasy of 'free' and 'spontaneous' markets, the commodity traded on carbon markets depends for its very existence on the on-going efforts of techno-scientific knowledge and political power. With the 1997 establishment of the Kyoto Protocol, and under the influence of economic expertise, climate change governance has increasingly come to rely on markets and their associated techniques of calculative agency. In the process, European and transnational trade in carbon dioxide equivalents has become an *explicit* matter of political design, generating new controversies and impacting upon an expanding range of distant

1 The poll can be visited here: http://rabble.ca/polls/carbon-trading-just-neoliberalism-under-green-veil. On the notion of opinion polls as productive political technologies, see Osborne and Rose (1999).

places, social groups and biophysical resources. Carbon markets, in short, arguably manifest one of the most ambitious political projects of neoliberalizing nature.

By tracing some of the particularities of how carbon has been forged as a neoliberal object – in-between the worlds of techno-science, economics, politics and public concern – this chapter engages discussions as to what 'actually existing neoliberalisms' imply in the science-dependent practices of transnational environmental governance. In particular, it picks up on the theoretical assertion, widely shared among Foucauldian governmentality approaches, that processes of neoliberalization entail relations of power and knowledge beyond the state, vested in specific assemblages of entangled global and local dynamics, and driven by the infiltration of market truths and calculative logics into domains of politics (Rose and Miller 1992, Ong 2006). Such power-knowledge dynamics create subjects and mould practices in particular ways, across diverse sites and spheres, with variable and sometimes unpredictable consequences.

This chapter aims to show that, in the case of carbon markets, the political technologies at work seek to mould a particular human agency, one that I will dub 'homo carbonomicus', the carbon- and price-calculative subject. Whether materialized in governmental offices, private companies, environmental NGOs or individual consumers, homo carbonomicus, I try to show, is an inherently *hybrid* subject, at once techno-scientific, economic and political (Callon 2009). Its calculative abilities are distributed within extended transnational networks that seek to render disparate concerns commensurable, across substantial differences in emerging political norms. Alongside feeding a new carbon economy of financial investors, consultants, project developers and green-tech firms, carbon markets and their associated calculative techniques thus emerge as important connecting points between the workings of everyday lives and the global-level political architectures attempting, but so far largely failing, to govern climate change.[2]

The hybridity of homo carbonomicus has a number of analytical implications for how to read this case of neoliberalizing nature; and, by implication, for theorizing market-infused environmental governance more generally. These implications, I suggest, push at the boundaries of governmentality approaches, suggesting the need for theoretical cross-fertilizations. First, as a techno-ethical subjectivity, the effects of homo carbonomicus cannot be delimited to commodification within new market spaces; rather, attempts to frame carbon as calculable market object constantly 'overflow' in the direction of techno-scientific controversies and unaccounted-for political concerns (Callon 1998, Lohmann 2005). Far from the *anti*-political hopes often invested in market-based regulation, such effects of *re*-politicization, I argue, become visible in work – inspired by Science and Technology Studies (STS) and actor-network theorist (ANT) Michel Callon – that

2 Space prevents me from expanding on the failures of climate governance, even as it underlies the positioning of this chapter. The fact that I write from Copenhagen, the city cast as 'Hopenhagen' and later 'Brokenhagen' during the UN COP15 conference in 2009, speaks to this issue of social embedding.

stress the *specific* design of calculative devices. Socio-technical specificities, this chapter argues, matter to the reading of neoliberal governmentalities of nature.

Second, in order to extend the analysis of hybrid carbon markets in ethico-political terms, the chapter turns to the sociology of critique and justification elaborated by French sociologists Luc Boltanski and Laurent Thévenot (2006). Beyond a monolithic market rationality, what this theoretical framework allows is to note the inherent plurality and divergence in political values (or 'orders of worth') articulated around carbon markets. Dominant elite pro-carbon market discourses, I argue, manifest a *fragile* compromise among market, industrial, and green forms of worth; a compromise that I dub the green neoliberal. This is the backdrop against which to understand why carbon markets are currently embroiled in serious political contestations around the world, placing them centre stage in collective hopes and frustrations manifested around hotly contested climatic futures. In the end, I suggest, homo carbonomicus needs to be re-embedded in a series of open-ended questions as to how civic solidarities may be extended across major global inequalities made increasingly manifest by global warming (Beck 2010).

Theoretical context: Neoliberal governmentalities of nature

As has been frequently noted, 'neoliberalism' is something of a rascal concept: at once pervasively employed, yet inconsistently defined and empirically imprecise, when diagnosing processes of market-oriented regulatory restructuring in post-1970s (trans-)national capitalisms (see e.g. Fourcade-Gourinchas and Babb 2002, Brenner, Peck and Theodore 2010). This poly-vocal situation pertains also to studies of nature's neoliberalization (Castree 2008a, 2008b), and to existing analyses of carbon marketization. Hence, while uniformly cast in this language, it remains disputed whether carbon markets are neoliberal by virtue of expressing an ideological market preference (Moore et al. 2011); by embodying a class project of atmospheric privatization (Lohmann 2010); by following the post-Marxist pattern of accumulation by dispossession (Bumpus and Liverman 2008); or by epitomizing a Foucauldian change from biopower to advanced (neo-) liberal government (Oels 2005).[3] Such ambiguities need attention since, as Castree notes (2008b: 158), 'neoliberalism' necessarily denotes phenomena that expand beyond specific cases like carbon markets.

Without entering into extended discussion, it is thus important to specify how and why this chapter adopts a largely (post-)Foucauldian rather than (post-)Marxist theoretical starting point. In overall methodological terms, governmentality

3 As part of expressing scepticism towards sweeping claims of millennial transformation, governmentality scholars often prefer to speak of 'advanced liberalism' rather than 'neoliberalism' (Brenner, Peck and Theodore 2010:199). To avoid confusion, however, I will only partly respect this terminological convention, preferring throughout the chapter to use terms like 'advanced (neo-)liberal government'.

approaches to neoliberalization tend to stay close to specific political settings, emphasizing the mundane practices, technologies of control and forms of subjectivity that underpin contemporary forms of market rule (Brenner, Peck and Theodore 2010: 199ff). Rather than grant systemic or globally hegemonic properties to a monolithic neoliberal ideology, governmentality scholars underline the hybridity and flexibility of market-oriented logics within changing regulatory landscapes. Based on this dynamic ontology, governmentality studies tend to foreground the productive capacities of neoliberalization, in terms of forging new 'responsibilized' state-subject relations within globalized assemblages of rule. Homo carbonomicus, this chapter suggests, exhibits exactly this translocalized and subjectified form of neoliberal governmentality, with carbon markets representing a specific trajectory of expert-driven market logics entering the global assemblage of climate change politics.

Based on such methodological premises, governmentality approaches resonate, I suggest, with what scholars in critical human geography – home to sustained discussions on nature's neoliberalization – identify as 'an emerging skepticism towards the coherence of neoliberalism as a programme of rule and the inevitability of fully formed neoliberal subjects' (Dowling 2010: 493; see Larner 2003, Castree 2008b). As stressed in studies of actually existing neoliberalizations, as opposed to its abstract ideological foundations, an adequate understanding of such processes require 'a systematic inquiry into their multifarious institutional forms, their developmental tendencies, their diverse socio-political effects, and their multiple contradictions' (Brenner and Theodore 2002: 353). This methodological commitment does not preclude the existence of layered path-dependencies and cross-jurisdictional interdependencies; rather, it implies that all neoliberalizations are *hybrid* from the outset, and that we need attention to their 'non-trivial differences' (Peck and Tickell 2002: 383). By extension, this chapter builds from the conviction that inquiring into the non-trivial specificities of carbon markets can serve to provide more general insights into the forms, effects and contradictions of neoliberal environmental governance.

For the purpose of analysing how homo carbonomicus is configured within emerging climate-political spaces of calculation, two overarching analytical topics of governmentality approaches to neoliberalization seem particularly salient. First, as with other instances of nature's neoliberalization, carbon markets need to be situated against the backdrop of how biopolitics has been extended, with the rise of environmental discourses and regulations since the 1960s, to encompass a 'green governmentality' directed at optimizing biophysical processes (Rutherford 2007). Importantly, this green governmentality is increasingly embedded within the production and governing of inter- and transnational spaces (Larner and Walters 2004). Second, in Foucauldian terms, the forging of homo carbonomicus arguably manifest a particular power-knowledge formation, tied to the rise of economics as an authoritative form of environmental expertise. I will briefly discuss these interrelated topics, highlighting how my theoretical approach articulates with, but also reframes, existing insights from governmentality studies.

In terms of green governmentalities, one reading of carbon markets sees these as expressive of a wider transition, manifesting roughly from the 1960s to the 1980s, whereby an environmentally expanded bio-power gradually gives way to a market-based regime of advanced (neo-)liberal government. This argument is made forcefully, for instance, by McCarthy and Prudham, as part of their wider observation that by necessity, 'neoliberalism is also an *environmental* project' (McCarthy and Prudham 2004: 277). In this context, green bio-power refers to extensions of the governmentalizing tendencies of bio-power further into the non-human realm, as embodied since the 1960s in new nation-state bureaucracies, programs and norms to scientifically administer nature's resources. According to McCarthy and Prudham, this early form of environmentalism 'had some success in resisting liberalism' (McCarthy and Prudham 2004: 278).[4] Concomitant with the Reagan-Thatcher years, however, neoliberal adherence to homo economicus led to sweeping assaults on such Keynesian-era environmental regulations. Environmental subjectivities sensitive to risks and scarcities, this analysis suggests, survive today only as resistance to neoliberal projects – or, in the largely assimilated ideological form of 'free market' environmentalism (McCarthy and Prudham 2004: 279).

An analytically similar, if perhaps less politically charged reading of governmentalities is offered by Oels (2005), within the confined space of globalized climate change governance. Embodied in the so-called 'flexible mechanisms' of carbon emissions trading in the Kyoto Protocol is a clear adherence to market-based technologies of agency, serving to foster 'responsible' and calculating subjects primarily at the level of UNFCCC member states.[5] This commitment to economics-centric advanced (neo-)liberal government contrasts, according to Oels' analysis, to the bio-political mission of planetary management, natural science expertise and technological fixes institutionalized, since the late 1980s, in the administrative space of the Intergovernmental Panel on Climate Change (IPCC). What we witness in global climate regulation, then, is a specific version of the wider transition from bio-power to advanced (neo-)liberal government. Similarly to McCarthy and Prudham (2004), Oels (2005: 198) suggests that this transition 'must be understood in the context of the global rise of neoliberalism in the late 1970s and 1980s'.

For my present purposes, Oels' analysis is particularly apposite in pointing to how neoliberal environmental regulation is increasingly taking place in international governance bodies, heavily influenced by technocratic forms of knowledge (Moore et al. 2011). In the climatic domain, resulting transnational

4 For data documenting this rise of a 'world environmental regime' since the 1960s, and the way it alters legitimate state identities and bureaucratic structures, see Meyer et al. (1997).

5 UNFCCC is acronym for the 1992 United Nations Framework Convention on Climate Change, the main international environmental treaty in the climatic domain, and legal backdrop to the Kyoto Protocol.

rules and mechanisms, like carbon markets, come to impact upon otherwise far-away communities. While this chapter aims for a more grounded account of carbon marketization, it builds from Oels' governmentality insight into transnational neoliberal rule-making as driven by specific elite spaces of techno-scientific and economic rationalities, with generative consequences for shaping political subjectivities and public resistances.

At the same time, however, the dichotomous contrast drawn up (in subtly different ways) by McCarthy and Prudham and Oels – between bio-politics and advanced (neo-)liberal government – itself needs to be questioned. To Oels (2005: 200), the contrast revolves around a change in knowledge practices, from natural science techniques of bio-political planetary surveillance (remote sensing, computer models) to those calculative techniques, of cost-benefit analysis and risk assessment, associated with advanced (neo-)liberal economics. While Oels is right in pointing to the importance of economic expertise in forging carbon markets – a topic to which I return in the next section – her contrast is still misleading, because it downplays the inherently hybrid character of homo carbonomicus, a subject of *intertwined* natural scientific and economic calculations. Inspired by work in science studies, I argue that natural science is integral, both to the way carbon is framed as a marketable object (e.g. MacKenzie 2009b), and to the way this framing ends up overflowing into new public controversies (Callon 1998).

A similar, if slightly more normative, argument should be directed at McCarthy and Prudham (2004), when they align the contrast between bio-politics and advanced (neo-)liberalism to a sense of *inherent* contradictions between environmentalist and market subjectivities. While this is a prominent trope of much 'green' neo-Marxist scholarship and environmentalist discourse since the 1970s,[6] as a theoretical *fait accompli*, this reading does not do justice to the inherent plurality and critical tensions manifest in spaces of carbon market politics. Notably, such tensions are visible in the way environmental NGOs around the world position themselves vis-à-vis emerging carbon market realities, exhibiting widely divergent ideologies and civic practices, from the highly market-enthusiastic to the highly market-critical (Blok 2011). Amongst transnational NGOs such as Greenpeace, for instance, the tendency is to engage carbon markets as open-ended political frameworks amenable to re-design along social and environmental lines, thus striking compromises between market, civic and environmentalist engagements.

More generally, as I will show in my analysis of the green neoliberal, carbon markets are inherently hybrid also in terms of the value commitments on which they rely. In this sense, neoliberal green governmentalities are spaces of ethical tension rather than monolithic rationalities. Beyond the cynical sense of corporate green wash and World Bank-style green capitalism that undergirds McCarthy and Prudham's 'free market' environmentalism, my sense of homo carbonomicus as embodying the green neoliberal is thus meant to give more analytical weight to ethico-political

6 One iconic expression of such commitments is the 1972 Club of Rome report on *Limits to Growth*.

ambiguities, frictions and hopes. This, I suggest, is what Boltanski and Thévenot's sociology of critique and justification allows us to analyse in more depth.

In sum, this section positions carbon markets as an important case study into wider processes of nature's neoliberalization. Here, the analytics of governmentality resonates, I argue, with the translocalized and subjectified shape of neoliberal market rule embodied in the figure of homo carbonomicus. By emphasizing the hybrid techno-scientific, economic and political character of carbon marketization processes, this analytics aim to take seriously some inherent tensions highlighted in existing work on green governmentalities. While neoliberal policies aiming explicitly at the non-human world are manifold, from bio-prospecting to ecotourism, fisheries quota trading and carbon offset schemes, environmental attachments *also* serve as important political sources of resistance to neoliberal projects (McCarthy and Prudham 2004, Castree 2008a). As the following analysis aims to show, the techno-ethical subject of homo carbonomicus finds itself inhabiting exactly such a space of manifold epistemic and political tensions.

Configuring homo carbonomicus: Neoliberal ideas in friction

By way of grounding carbon markets in specific spaces of transnational policy circulation, let me start by invoking a contextualized moment in my own ethnography of this neoliberal trajectory (see Greenhouse 2010).[7] Every year, environmental elites from around the world – governmental ministers, business leaders, natural scientists, NGO executives – meet for the Delhi Sustainable Development Summit (DSDS), to discuss global challenges in the midst of this fast-industrializing economy of the 'global South'. Under the auspices of IPCC Chairman Rajendra Pachauri, the 2008 meeting situated the event in the path 'from Bali to Copenhagen'.[8] Spanning a range of concerns, from local solar lantern charity efforts to devising an equitable post-Kyoto climate regime for the world, the meta-language of economics would flow effortlessly through the meeting. Like incantations, speakers kept repeating the tropes of 'working towards a low-carbon economy', 'provide incentives for consumers' and 'putting a global price on carbon'.

One speaker particularly wedded to the promise of carbon markets was Michael Walsh, Executive Vice President of the Chicago Climate Exchange (CCX), so

7 I attended the Delhi Sustainable Development Summit (DSDS) in February 2008, on invitation as 'observer'. I wish to thank the organizers (TERI) for making this valuable opportunity possible, and for generally committing (in however nascent form) to wider public engagement and accountability.

8 Like other expert-dominated domains, the world of climatic professionals is brimming with obscure technocratic insiders' language. From 'Bali' (2007) to 'Copenhagen' (2009) is a compressed way of talking about international climate negotiations, at this point (2008) invested with considerable political hopes.

far North America's main voluntary integrated carbon trading platform. Carbon markets, Walsh suggested, would 'drive capital to its highest environmental protection capacity', as evidenced already by the viable and business-friendly cap-and-trade system put in place with the European Emissions Trading Scheme (EU-ETS). In delivering this message, Walsh made ample use of a standard pro-carbon market narrative, referencing the 'success story' of how sulphur dioxide (SO_2) emissions trading made American acid rain counter-policies much more cost-efficient than initially projected by the authorities. Underlying this narrative is the notion that specifically *American* market experiences should be translatable to the *global* domain of climate regulation; a power-laden US position that shaped also the fragile international compromises struck in the Kyoto Protocol (Bulkeley 2001: 439).

Pointing to CCX's involvement with the growing market for Clean Development Mechanism (CDM) projects in India, Walsh concretized the sense that 'there is profit to be made from eco-efficiency'. Indeed, the Summit blended into a CDM market bazaar, with local businessmen striking new deals with international partners, and leaflets advertising the carbon credit rates of renewable energy projects in India. As a projects-based market, this is how the CDM works: the Kyoto Protocol invites 'Northern' governments and businesses to invest in emission reductions in the global South, and to transfer the 'surplus credits' to their own carbon accounts in order to meet reduction targets (Lohmann 2005). New carbon professionals in project management and verification help forge the deals; the CCX, on their part, invests in methane capture projects from landfills in Mumbai, selling credits on to Western companies. Such connections have brought more political and commercial actors in places like India, China and Brazil into the 6.5 billion US dollars CDM market.[9] The Delhi Summit was one venue of market extension.

In his talk, Walsh went further in the attempt to make palpable the promises of carbon markets to his Delhi audience of eco-elites; notably, he came close to providing the abstraction of homo carbonomicus with a personalized face. Showing a picture of a young white American girl in front of the Chicago Stock Exchange in 1994, Walsh explained that this 12-year-old had walked into the marketplace one day and bought 3.000 US dollars' worth of SO_2 permits, taking them off the market to raise costs on the part of air-polluting industries. Heralding these 'brave actions', Walsh drew the general conclusion that carbon markets 'need to be simple enough for the public to access them'. In a striking rhetorical move, the economics of carbon markets was thus aligned to public expressions of environmental concern – prefiguring, we should add, how carbon markets have become spaces of green political consumption. The Chicago girl, in short, is one embodiment of homo carbonomicus, a hybrid techno-ethical figure, American style.

9 This figure is from 2008, corresponding to the year of the DSDS. As a market volatile to the global economic crisis, the CDM contracted considerably in 2009 (World Bank 2010: 1); it is now growing again.

I reference these ethnographic moments to conjure a counter-image to the way neoliberal ideologies are sometimes portrayed as a unified hegemonic order, diffusing smoothly across the globe, country by country (see Ong 2006). Instead, carbon markets illustrate the more general claim that market-oriented technologies of governance emerge and spread within uneven and bounded networks, where cadres of expert 'technopols' make up transnational circuits of ideational persuasion and experimentation (Brenner, Peck and Theodore 2010: 214ff). Climatic elite networks, like the DSDS, are shot through on all sides by power, inequalities and exclusions, manifested in part through hierarchies of authoritative expertise. As such, they operate as spaces of friction, in which actors 'otherwise engaged' – US carbon trading professionals, Indian charity entrepreneurs, environmentalists, concerned scientists – meet across differences in interests and values (Tsing 2005, Harvey and Knox 2008). Far from the cold abstractions of textbook economics, carbon markets are spaces of vital and often contradictory energies.

Along such lines, we may trace the gradual configuration of homo carbonomicus as a political innovation, from the early development of emissions trading models in the US Environmental Protection Agency (EPA) of the late 1970s, up until the regime formation of carbon markets in and beyond the UN Kyoto Protocol in the late 1990s (Voss 2007). From the gestation period, questions of technical market design have been closely tied to shifting political coalitions, as regulators sought to use emissions trading as means of reconciling economic and environmental interests. Via prototype testing in acid rain regulations, emissions trading gradually gained widespread elite acceptance as an environmental policy instrument in the United States. While some environmental NGOs confronted the EPA in court, claiming the ethical unacceptability of putting pollution on sale, other groups, such as the Environmental Defense Fund (EDF), was directly involved in innovating the market model. As noted, techno-ethical divergences continue to shape environmental NGO practices vis-à-vis carbon markets (Blok 2011).

The key moment in the social life of homo carbonomicus, arguably, is the fragile compromise reached among divergent political views, notably those of 'North' and 'South', during the Kyoto negotiations (Bulkeley 2001: 439). Against the resistance of the European Union (EU), US diplomats enthusiastic about emissions trading succeeded in striking a political deal with countries of the global South, led by Brazil, by committing to technology and financial transfers through the new CDM framework (MacKenzie 2009a). Once again, as the US-led ideational transfer went transnational, technical questions of market design was thus caught up amidst divergent tensions and hopes, spurred partly by international business interests, science-based environmental concerns, and divergent views on the proper global distribution of climatic burdens and responsibilities. In this case, one main effect was a market, the CDM, which looks very different from textbook emission markets (Callon 2009) – and whose mechanisms of value-generation continue to spur ethico-political controversies, both 'South' and 'North'.

Somewhat ironically, with the disavowing of Kyoto by the US government, the EU Commission became, from the early 2000s, the main institutional hub of carbon markets, reframing an instrument that it had previously contested internationally. Spurred by US economic experts flying in as consultants, European policy networks took up the hype of carbon markets; and from 2005 onwards, the EU-ETS has established a market for 2.2 billion tons of carbon allowances, from 11,500 large-scale industrial installations, and with a daily transaction volume of 60 million Euros (Voss 2007: 339).[10] The underlying principle of this mandatory cap-and-trade system is simple enough: acting as regulator, the Commission sets a Europe-wide 'cap' – a maximum allowable aggregate quantity of emissions – and distributes the corresponding allowances to high-emitting companies. These companies may then 'trade' allowances among themselves, according to their financial incentives for either buying more allowances on the carbon market (to comply with regulations), or invest in emission reductions and sell surplus allowances.

In practice, every step of designing the EU-ETS involves contentious negotiations: where to set cap levels for different countries; how to distribute allowances (free or via auctioning); which sectors to include (aviation to be included in 2012); how to monitor fraud; and so on. Gradually, within this new market configuration, multiple agencies come to reinforce each other and add momentum to homo carbonomicus: alongside a carbon industry of traders, exchanges and project developers, the EU-ETS depends on multi-level infrastructures of public agencies, emission auditors, think tanks, law firms, consultancies, and so on. By linking up with the CDM, moreover, the EU-ETS enlists the techno-scientific apparatus of the IPCC, needed to translate greenhouse gasses into 'carbon equivalences' (Callon 2009). This is how, for instance, the CO_2 emissions generated by a British university can be made equivalent, in biophysical and market terms, to HFC-23 industrial gas decomposition in China (MacKenzie 2009b).

Like the CDM, the EU-ETS continues to lead a turbulent political life, under attack from various sides – including from business constituencies that see cap-and-trade mainly as imposing new costs, hampering their global competitiveness. Meanwhile, ethico-political realignments are clearly discernible: green-tech companies welcome carbon markets as spurs to new technological innovation; and high-profile environmental groups, like the World Wide Fund for Nature (WWF), has gone from moral critique to active engagement with market rule-setting. In this sense, carbon markets implicate and spur new ambiguities, as environmentalist attitudes towards market-inspired policies gradually change: while most groups favour eco-taxes to cap-and-trade, the compliancy mechanisms of the latter still make it preferable to voluntary agreements with industry (Bomberg 2007: 254). Far from a subject of affection, many European environmentalists thus nowadays tend to cautiously accept homo carbonomicus (Blok 2011).

10 These are all 2006 figures. According to World Bank data, by 2009 the EU-ETS volume had grown to 6.3 billion tons of carbon, with an annual transaction value of app. 118 billion US dollars, making this by far the largest existing carbon market in the world in terms of both volume and value (World Bank 2010: 1).

To sum up, the main argument here is that, embodying a green neoliberal compromise, homo carbonomicus has gradually taken on a social life of its own, circulating the world in powerful expert networks, thereby weaving trajectories of institutional restructuring. As technologies of governance widely promoted by transnational climatic elites, carbon markets exemplify the wider tendency towards technocratic politics associated with neoliberalizations in international spaces (Moore et al. 2011). Still, I argue, such technocratic spaces are shot through with frictions. Whereas enthusiastic economizers in the Delhi summit would dream of putting a *global* price on carbon, market realities remain fragmented, with *different* market designs (EU-ETS, CDM, voluntary markets) operating across national and regional scales. Likewise, the exact contribution of these markets to a 'low-carbon economy' remains contested, amidst a generalized sense of failure in addressing the challenges outlined by science. Following these suggestions, I turn next to explore how different carbon market designs overflow their own spaces of calculability and accountability, generating new public dynamics of ethico-political contestation.

Market overflows: Dynamics of public contestation

Given that global warming has been memorably described by economist Nicolas Stern as 'the biggest market failure the world has ever seen', the ironies involved in addressing the problem by generating *more* (carbon) markets is not lost among environmentalists influenced by 'leftwing hostility to an extension of market relations' (MacKenzie 2009b: 449). Analytically, however, debate among pro- and anti-market forces in the climatic context suffers from a general tendency, equally pronounced on both sides, of reifying 'the market' as an invariant economic structure, distributing hopes and failures along pre-set lines. This tendency, I suggest, blurs the inherent hybridity of homo carbonomicus as a techno-ethical subject, by purifying the economic realm from its constitutive entanglement in techno-scientific and political concerns. To see how marketization comes to have both anti-political *and* political effects (Barry 2002), we need instead to explore the *different* calculative techniques and accountabilities embedded in carbon market designs. Further, we need to explore how such designs engender new ethico-political spaces of civic engagement, thereby overflowing their own frame of calculability (Callon 1998).

As a first approximation, the list of entities routinely overflowing current carbon market designs seem substantial: it includes the majority of the world's greenhouse gas emissions,[11] as well as a range of ethico-scientific concerns expressed by environmental NGOs, low-lying island states, and communities in

11 Even with cap-and-trade-like mechanisms under development in high-emitting countries (Japan, Australia, US), on a global scale, the total share of greenhouse gasses (CO_2 equivalent) covered by *any* kind of carbon market remain below 20% (author's own

the global South (Blok 2010). This situation is compounded by the fact that carbon markets constitute only one amongst plural socio-political measures – of legal target-setting, technological innovation, developing countries' right to resources, urban planning, and so on – currently grappled with vis-à-vis the unruly issue of global warming (Callon 2009). Frictions emerge at the overlapping boundaries of specialized concerns, as climatic actors seek to carve out the political norms needed to delimit the proper role of markets. This is visible, for instance, in on-going attempts to include tropical forest protection in the CDM market, attempts that meets with considerable resistances amidst long-standing global controversies (Tsing 2005). Here, Brazilian and Indonesian forests come to constitute some of the conflict-ridden frontier zones for the as-yet unsettled future of homo carbonomicus.

In a more fundamental sense, overflows in the direction of unaccounted-for public concerns seem to follow closely from the sheer range of techno-scientific and political ambiguities built into current carbon market designs. While the science of climate change has achieved considerable sophistication in recent years, uncertainties still abound on core issues of modelling the carbon cycle; determining the range of sources and sinks; measuring the carbon fluxes associated with market mechanisms; and determining long-term equilibrium targets for atmospheric CO_2 concentrations (Frame 2011). Such 'global' techno-scientific interpretive flexibility spills into the specific 'local' calculative techniques used to frame marketable carbon sinks in the CDM market, where methods and rules of technical accountability remains widely contested (Blok 2010). Meanwhile, in political terms, the Kyoto Protocol avoids such ambiguities largely by conjuring and distributing available carbon sinks roughly in proportion to *existing* levels of developed countries' greenhouse gas emissions – levels that are, however, widely acknowledged to be *unsustainable* in terms of both short- and long-term environmental futures (Lohmann 2005).

Analytically, what these techno-scientific and political concerns show is that any construction of a carbon economy involves a process of 'norm-isation', as market action is inextricably entangled in ethico-political groundings of what is 'right', 'good' and 'better' in terms of reacting to global warming (Goodman and Boyd 2011: 103). However neoliberal by design, carbon markets thus cannot be disentangled from basic public-political debates on responsibility, accountability and environmental care. Indeed, as hybrid techno-ethical objects, issues of care and responsibility for Others – including distant strangers and future generations – is embedded in the very price calculations performed over carbon emissions and offsets (Goodman and Boyd 2011: 106). Unlike its predecessor homo economicus, which managed for a long time to sustain the fantasy of a 'pure' market calculability, homo carbonomicus explicitly quantifies across *divergent* values and attachments. The fact that such operations of impure calculation

calculation, based on World Bank and UNEP data). In Europe, this share sits around 40% (according to EU data). All such figures are contested.

(or 'qualculation') will often be inconsistent, uncertain, and partly obscure to the calculating agencies themselves, however, is no barrier to their practical working-out in specific contexts (Callon and Law 2005).

Processes of impure calculation, and its public contestation, may be observed across various sites and scales of market extension. They are visible, for instance, in the way carbon markets become spaces of individualized ethico-political consumption, opening up new subjective ambiguities. On the one hand, voluntary offset markets for air travel and other consumption practices are widely dismissed as little more than modern-day middle-class indulgencies. Amongst participants in low-carbon living experiments, for instance, carbon offsetting is critiqued in reference to *real* instances of green citizenship (Marres 2008). On the other hand, the practice of individuals buying up and destroying carbon credits on the EU-ETS market, to raise costs on high-polluting industries, has spread in tandem with the market itself, in part by being promoted by environmental NGOs and renewable energy provision companies. In this sense, divergent sites of homo carbonomicus lend themselves to different emerging norms of il-/legitimate ways of expressing individual ethico-political concerns through market mechanisms.

Similarly, at the level of national climatic policies, the fact that state agencies involve themselves in new CDM market arrangements, as part of complying with international Kyoto obligations, lends itself to multiple lines of public criticism. Such dynamics of re-politicizing market arrangements have been most visible in protracted public controversies surrounding the large-scale industrial gas (HFC-23) decomposition projects in China, financed via the CDM by European governments and businesses.[12] By interrogating the calculative techniques underlying the trade, and the way it effectively subsidizes a heavily climate- and ozone-damaging chemical industry, environmental NGOs here succeed in raising critical public debates as to the (un-)accountabilities of national government policies. Such debates easily spill into a wider sense of public scepticism with inadequate governmental action, as buying up CDM credits from far-away projects is discursively pitted against the *real* work of de-carbonizing the domestic economy. Once again, carbon markets are thus re-entangled in techno-scientific and political contestations, this time related to emerging commitments to long-term national low-carbon transitions.

Finally, dynamics of impure calculation become potentially explosive, in ethico-political terms, when publicly re-entangled in those dense webs of global socio-economic inequalities that carbon markets like the CDM span. Spurred by critical NGO attention to the global injustices of specific carbon market connections, as documented via alternative techniques of ethico-political witnessing, such passionately enraged overflows travel along with homo

12 The following observations are based mostly on personal familiarity with Danish climate politics. Similar critical CDM debates have occurred in other European countries, including the UK (Blok 2011).

carbonomicus, occasionally making it into wider public debate (Blok 2010).[13] This happened, for instance, when the UK-based carbon offset trade involving the transfer of human-powered water pumps to northern India came under public scrutiny, in both the UK and Indian media, as instantiating a neo-colonial sense of India's rural poor carrying the 'white man's burden' of caring for the global climate (see Sethi 2007). In such moments, new world-spanning carbon market connections – including new NGO witnessing networks – serve inadvertently to bring public visibility in the 'North' to various 'Southern' concerns with climate change, livelihoods and resources. In the process, more cosmopolitan visions of climatic justice may slowly gain ground (Beck 2010).

In sum, as hybrid objects of impure calculations, carbon marketization has both anti-political *and* (re-)political effects, contracting and expanding new ethico-political spaces of civic contestation (Barry 2002). This is an important observation, because civic engagement serves as counterpoint to the powerful technocratic spaces of climatic elites propagating carbon markets. More generally, in pointing to the multiple lines of situated political contestation, this analysis attempts to question the idea of 'post-politics' often associated with nature's neoliberalization (e.g. Swyngedouw 2010). As such, it resonates rather with the notion of 'sub-politics': as carbon markets forge new contentious connections, economic, techno-scientific, community and consumption practices, so far considered un-political, may gradually emerge as contestable and experimental sites of nascent climatic citizenships (see Bulkeley 2001). This expansion in (sub-)political forums (Callon 2009), however, in no way guarantees the forging of collective powers sufficient to address global warming; in this sense, inside and outside of carbon markets, world climatic risks still seems largely a case of organized irresponsibility (Beck 1998).

The green neoliberal: Justification, critique, compromise

The fact that carbon markets get re-entangled in diverse sub-political controversies, across a plurality of sites and scales, in itself does little to address the suspicion that such entanglements may represent only superficial glitches on the path to neoliberal market rule. In the end, to see why carbon trading is not simply (as the Canadian pollsters had it) 'neoliberalism under a green veil', we need to complicate the normative analytics of market calculation at deeper levels, by articulating the inherent *diversity* of values manifest in any market-infiltrated social ordering. Invoking the value-pluralist sociology of justification and critique elaborated by Luc Boltanski and Laurent Thévenot (2006), in this final section, I

---

13 One creative illustration of such alternative techniques of public witnessing is the carbon trade-critical documentary movie, aptly entitled *The Carbon Connection*, made by the NGO Carbon Trade Watch. The movie may be watched here: http://www.carbontradewatch.org/carbon-connection/index.html.

aim to elaborate why carbon markets simply cannot exist on the basis of *market* commitments alone.[14] Instead, I suggest, they represent fragile compromises between market, industrial, and green orders of worth. This goes a long way in explaining significant cross-cultural differences in their uptake and legitimacy.

In a series of case studies into public environmental controversies, Thévenot (and co-authors) document the variety of ways in which 'natural' entities can be qualified for cognitive-moral evaluation in contemporary Western societies (Thévenot, Moody and Lafaye 2000, Thévenot 2002). Following a general model of the ordinary sense of justice operative in situations of conflict (Boltanski and Thévenot 2000), they show how environmental disputes revolve around a limited plurality of moral grammars, or 'orders of worth', available to actors' justifications and critiques. The basic model recognizes six orders of worth in political modernity (Wagner 2008): entities, including 'nature', may be valued as part of short-term monetary exchanges (market); in terms of long-term planning and technical efficiency (industrial); as expressing collective welfare and solidarity (civic); as integral to local heritage and customs (domestic); as affording passion and experiences of sublimity (inspired); and as signs of fame and popularity (opinion). With the rise of environmentalism, moreover, entities may also be qualified according to 'green' criteria, of environmental friendliness, ecosystem health and planetary sustainability.[15]

As grammatical repertoires of evaluation, orders of worth are not tied to discrete domains or institutions of society; rather, they are available to actors across diverse situations of coordination and dispute. Nevertheless, empirical-historical inquiry will reveal differential patterns of political culture, according to institutionalized constraints and commitments. Comparing environmental disputes in France and the United States during the 1990s, for instance, Thévenot (and co-authors) show how, amidst value diversity, market evaluations were much more frequent in US debates, often in combination with 'green' and 'civic' arguments (Thévenot, Moody and Lafaye 2000: 263f). This kind of evaluative compromise – where market access comes to be associated with equal rights to the environment – in many ways parallel the culturally peculiar version of homo carbonomicus enacted by Michael Walsh in Delhi. In the French context, Thévenot's study suggests, such associations will likely seem misplaced. Instead, French debates tend to associate 'green' worth to a coupling of expert planning competency with civic equality, testifying to legacies of an engineering-heavy technocratic state apparatus.

justice

14 To be clear, I claim little originality for this observation on 'the embeddedness of markets'; it is shared widely in socio-political thinking, including in Foucauldian and Marxist theorizing. What *is* original, I believe, is the attempt to articulate, via Boltanski and Thévenot, what *other* sources of ethico-political commitments sustain specific market orders. This, I suggest, is seldom made explicit in socio-political theorizing.

15 The rise of this so-called 'green city' (*cité verte*) has been extensively debated in French social science during the 1990s. See Lafaye and Thévenot (1993) for the opening formulation of the questions.

Such long-term cultural-political patterns of moral evaluation, I suggest, are important for understanding the divergent inflections given to carbon markets, as hybrid techno-ethical objects, across different social contexts. Hence, as Goodman and Boyd point out, 'even the magic of the (carbon) market is now too much regulation' in the US 'hotbed of neoliberal capitalism' (Goodman and Boyd 2011: 102); and this contrasts sharply, as noted, to how homo carbonomicus has attained institutional stabilization in the EU. Whereas US debates tend to evaluate carbon markets mostly for their short-term monetary effects (in both critiques and justifications), European evaluations tend towards a much stronger emphasis on these markets as part of a long-term planned future. Arguably, this emphasis is characteristic of the 'industrial' criteria of evaluation strongly embedded in technocratic EU policy-making. Industrial worth, in this sense, inscribes carbon markets into climate change evaluative frameworks based on the efficiency of public investments, techno-scientific competency, and long-term planned growth.

What this means is that, in the European context, public (non-)confidence in carbon markets is inseparably tied up with wider negotiations of faith in the willingness of policy elites to tighten environmental standards when setting long-term 'industrial' planning targets. This effect is nowhere more obvious than in critiques of the 'over-allocation' of carbon allowances that led to a virtual collapse in market prices during the first phase (2005–2007) of the EU-ETS (MacKenzie 2009a). For the second phase (2008–2012), the Commission took steps to tighten overall allocation levels against reluctant nation-states; at the same time, moves were made towards auctioning rather than free allocation of allowances. What this shows, first of all, is that carbon marketization is an experimental process of trial-and-error – a process led by technocratic policy elites, but to some extend responsive to dynamics of public criticism (Callon 2009). Moreover, in the language of Boltanski and Thévenot (2006), it shows that basic carbon market mechanisms of supply and demand are essentially engineered according to 'industrial' criteria, in the process embedding social compromises over environmental ('green') concerns.

This is also the analytical context in which to understand changing European NGO evaluations vis-à-vis carbon markets, as critical denunciations give way in many cases to cautious acceptance and engagements with standard-setting (Blok 2010). Such work is visible, for instance, in how the influential Climate Action Network (CAN) umbrella organization has been elaborating, since the early 2000s, its own set of justificatory criteria for critically supporting carbon markets as a green neoliberal compromise. Apart from thus accepting that, under specific conditions, 'market' and 'green' criteria of evaluation may function together, CAN stresses that political targets (the caps of cap-and-trade) must be 'credible and ambitious', and that the decision-making process should be 'transparent and open to participation' (quoted in Bomberg 2007: 252). In doing so, CAN adds further layers of ethical qualifications to the situation, by stressing the importance of both 'industrial' (ambitious targets) and 'civic' (open participation) commitments. By calling for 'ambitious' market standards, environmental NGOs carve out an ethico-political space in which policies can be measured – and publicly critiqued – against self-defined criteria of what should constitute an environmentally credible ('green') commitment.

Similar dynamics, whereby environmental NGOs achieve standard-setting powers usually associated with expert regulatory agencies, have been more pronounced in the transnational domain of the CDM. Hence, arguably the most influential NGO engagement with carbon markets so far has been the WWF-initiated gold standard accreditation rules for 'high quality' credits in the CDM market, a standard that has effectively split the market in two according to shades of green worth. Unlike the formal UN-backed rules, the gold standard excludes the most environmentally dubious credits, covering mainly renewable energy technology projects; moreover, it seeks to assure that sustainability benefits accrue also to local communities in the global South (MacKenzie 2009b: 451). In this sense, the gold standard scheme seeks to both strengthen the parameters of environmental sustainability ('green' worth), and to add 'civic' concerns with global socio-economic distribution, within the space of ethico-political friction constituted by the market frame. By now, the standard is backed by more than 60 environmental NGOs around the world, and while accreditation remains voluntary, it proves attractive to companies and governments sensitive to their public 'green' reputation (Blok 2011).

To sum up, the overall argument here is that, as an impure neoliberal subject, homo carbonomicus emerges as a fragile ethico-political compromise in spaces of inherent value pluralism, tension and critique. Extending the analytics of Boltanski and Thévenot (2006), I argue that carbon markets depend for their public-political justification on associating short-term monetary incentives (market worth), long-term expert planning (industrial worth), and notions of environmental sustainability (green worth). In this constellation, parameters of green worth arguably remain in flux, in both scientific and public discourse, amidst on-going controversies on how to work towards long-term climatic sustainability. Still, however neoliberal, carbon markets cannot sustain themselves *purely* by market commitments – something made evident in cross-cultural comparison among Europe and the US. At the same time, the language of Boltanski and Thévenot also helps specify important concerns excluded or downplayed by carbon market arrangements; not least in terms of civic concerns with North-South socio-economic divides. Here, environmental and other NGOs may play important democratic roles in critiquing market-infused injustices and elaborating more cosmopolitan forms of civic solidarity (see Beck 2010).

Concluding remarks: Ethicizing/politicizing the neoliberal?

This chapter traces the making of carbon as a techno-scientific and neoliberal object, across a variegated regulatory landscape of fraught attempts to govern climate change at national, regional and global scales. Since the 1990s, carbon markets have emerged as an ambitious political project of nature's neoliberalizing, involving the large-scale construction, from scratch, of a new carbon economy. Drawing on the (post-)Foucauldian analytics of green governmentality, the chapter shows how carbon markets arise from power-knowledge relations beyond the state, driven by networks of technocratic expertise serving to fuse market truths and calculative logics into global environmental politics. These power-knowledge dynamics mould and configure particular human practices, giving rise to homo carbonomicus, the carbon- and price-calculative subject. The hybrid and impure calculative techniques associated with homo carbonomicus, I argue, has come to constitute important points of connection between everyday lives and global-level political architectures, across diverse sites and scales of ethico-political ambiguities.

While adopting methodological starting points compatible with governmentality studies, the chapter also extends this approach via two theoretical cross-fertilizations, serving to reconfigure homo carbonomicus as a less fully formed and more hybrid neoliberal subject (see Dowling 2010). First, work in Science and Technology Studies (STS) serves to emphasize how specific carbon market designs emerge from the conjoint efforts of economic, techno-scientific and political actors, devices and sites (Lohmann 2005, Callon 2009, MacKenzie 2009b). Carbon market construction, on this view, is a site of (sub-)political engagement, where economic experts, regulators, business lobbyists, IPCC scientists and environmental NGOs all play roles in testing and reworking specific market designs. Like most inter- and transnational neoliberal governance spaces, carbon market design remains for the most part technocratic and elitist (see Moore et al. 2011). At the same time, however, I emphasize that techno-scientific and political ambiguities make carbon markets overflow, opening up new public spaces of contestation.

This leads to the second strand of theoretical reconfiguring, aiming to qualify the normative underpinnings of governmentality studies. What Boltanski and Thévenot's sociology of justification and critique allows, I suggest, is to show how – for all its neoliberal attachment to market rule – homo carbonomicus remains a fragile compromise among plural ethico-political concerns. Hence, the specific 'norm-isation' of cap-and-trade carbon markets includes a strong commitment to long-term technocratic expert planning, in large part explaining why the European Union has emerged as institutional hub of regulatory experimentation. Meanwhile, the analytics of Boltanski and Thévenot help specify the moral grammars available to environmental NGOs and other engaged civic-political actors, in trying to tighten the questionable environmental credentials (green worth) of the CDM market. By making North-South socio-economic divides visible via critical NGO witnesses, specific CDM projects may inadvertently come to spur the nascent emergence and spread of more cosmopolitan calls for climate justice (Blok 2011).

The combined effect of these re-specifications, I believe, is a theoretical approach to neoliberalization processes that may have implications well beyond the specific case of carbon markets, reaching into wider questions of techno-science and neoliberal environmental governance. First, the approach advocated in this chapter serves as an antidote to unfortunate habits of thought – widespread, I claim, in both liberal economic and (some strands of) Marxist scholarship – that tend to reify 'the market' as an unchanging, trans-cultural, structural formation. Quite to the contrary, as STS scholars have shown (e.g. Knorr Cetina and Preda 2001), the rules of *divergent* market transactions are increasingly the product of applied economic knowledge, reliant on heterogeneous data, observation, and validation techniques. In this process, markets are opening up to hybrid forms of knowledge, both techno-scientific and environmentalist. If the case of carbon markets is anything to go by in this respect, *all* markets will increasingly have to be thought of as *explicit* objects of techno-scientific and political design, subject to new accountabilities vis-à-vis the epochal ecological problems facing humanity (Callon 2009).

This increasing hybridization of markets, techno-science and politics in no way entails that particular market designs, inside or outside the environmental domain, are necessarily benign, unproblematic, or in any sense adequate to the problems at hand. Indeed, I have suggested that – amidst more generalized irresponsibility and political failures – current carbon markets are highly inadequate to the task of addressing the concerns raised by global warming. Nevertheless, as a second implication, this still suggests a different theoretical approach to the (sub-)politics of markets. Caught between 'the economy' (liberal) and 'political economy' (Marxist), I argue, markets are too often depicted as either autonomous from politics or as entirely subordinate to political ideologies. The alternative approach advocated here suggests a more variable, contingent and context-sensitive analytics, which recognizes both the anti-political *and* the political effects of neoliberal marketization projects (Barry 2002). While markets may be deployed along with technocratic dreams of post-politics, through continuous overflows, they end up generating spaces of civic contestation, tied to novel ethico-political subjectivities.

This brings me to a third and final theoretical implication of the present study: with Boltanski and Thévenot (2006), I hope to have suggested, using the case of carbon markets, the need for elaborating a 'sociology of critique' alongside the more standard 'critical sociology' in analyses of actually existing neoliberalizations. As Castree points out (2008b:132), the bulk of research on nature's neoliberalization is 'broadly unsympathetic to the project of "market rule"'.[16] While this may certainly be warranted (in and beyond carbon markets), what is problematic, I believe, is the way in which the rise of neoliberalism, as Peter Wagner notes (2008:84), has been too easily 'accepted by critical thinkers as a hegemonic *pensée unique* to which "there is no alternative" in an era of globalization'. Besides misconstruing the normative infrastructure of societies, this kind of analytics serves only to constrain the political imagination.

16 Castree makes his remark in the context of human geography, but it could well be extended to sociology.

What remains to be done, instead, is to re-embed critiques of neoliberal market rule, and their corresponding justifications of alternative orderings, in dense layers of everyday public sense-making in value-pluralist societies, including among environmental NGOs. While the climatic citizenships of the future are only yet visible in nascent form, carbon markets may be construed as an ethico-political experiment, reworking the boundaries of markets, techno-science and public action around new moral configurations. At its critical edges, and beyond dominant market-industrial attachments, homo carbonomicus may still prove an agent of new cosmopolitan solidarities, much needed if current failures are to turn into new hopes of greener futures.

References

Barry, A. 2002. The anti-political economy. *Economy and Society*, 31(2), 268–84.

Beck, U. 1998. *World Risk Society*. Cambridge: Polity Press.

Beck, U. 2010. Remapping social inequalities in an age of climate change: For a cosmopolitan renewal of sociology. *Global Networks*, 10(2), 165–81.

Blok, A. 2010. Topologies of climate change: Actor-network theory, relational-scalar analytics, and carbon-market overflows. *Environment and Planning D*, 28(5), 896–912.

Blok, A. 2011. Clash of the eco-sciences: Carbon marketization, environmental NGOs, and performativity as politics. *Economy and Society*, 40(3), 451–76.

Boltanski, L. and Thévenot, L. 2000. The reality of moral expectations: A sociology of situated judgment. *Philosophical Explorations*, 3(3), 208–31.

Boltanski, L. and Thévenot, L. 2006. *On Justification: Economies of Worth*. Princeton, NJ: Princeton University Press.

Bomberg, E. 2007. Policy learning in an enlarged European Union: Environmental NGOs and new policy instruments. *Journal of European Public Policy*, 14(2), 248–68.

Brenner, N. and Theodore, N. 2002. Cities and the geographies of 'actually existing neoliberalism'. *Antipode*, 34(3), 349–79.

Brenner, N., Peck, J. and Theodore, N. 2010. Variegated neoliberalization: Geographies, modalities, pathways. *Global Networks*, 10(2), 182–222.

Bulkeley, H. 2001. Governing climate change: The politics of risk society? *Transactions of the Institute of British Geographers*, 26(4), 430–47.

Bumpus, A. and Liverman, D. 2008. Accumulation by decarbonization and the governance of carbon offsets. *Economic Geography*, 84(2), 127–55.

Callon, M. 1998. An essay on framing and overflowing: Economic externalities revisited by sociology, in *Laws of the Markets*, edited by M. Callon. London: Blackwell, 244–69.

Callon, M. 2009. Civilizing markets: Carbon trading between *in vitro* and *in vivo* experiments. *Accounting, Organizations and Society*, 34(3–4), 535–48.

Callon, M. and Law, J. 2005. On qualculation, agency, and otherness. *Environment and Planning D*, 23(5), 717–33.

Castree, N. 2008a. Neoliberalising nature: The logics of deregulation and reregulation. *Environment and Planning A*, 40(1), 131–52.

Castree, N. 2008b. Neoliberalising nature: Processes, effects, and evaluations. *Environment and Planning A*, 40(1), 153–73.

Dowling, R. 2010. Geographies of identity: Climate change, governmentality and activism. *Progress in Human Geography*, 34(4), 488–95.

Fourcade-Gourinchas, M. and Babb, S. 2002. The rebirth of the liberal creed: Paths to neoliberalism in four countries. *American Journal of Sociology*, 108(3), 533–79.

Frame, D. 2011. The problems of markets: Science, norms and the commodification of carbon. *Geographical Journal*, 177(2), 138–48.

Goodman, M. and Boyd, E. 2011. A social life for carbon? Commodification, markets and care. *Geographical Journal*, 177(2), 102–9.

Greenhouse, C. 2010. Introduction, in *Ethnographies of Neoliberalism*, edited by C. Greenhouse. Philadelphia: University of Pennsylvania Press, 1–10.

Harvey, P. and Knox, H. 2008. 'Otherwise engaged': Culture, deviance and the quest for connectivity through road construction. *Journal of Cultural Economy*, 1(1), 79–92.

Knorr-Cetina, K. and Preda, A. 2001. The epistemization of economic transactions. *Current Sociology*, 49(4), 27–44.

Lafaye, C. and Thévenot, L. 1993. Une justification écologique? Conflits dans l'aménagement de la nature. *Revue Francaise de Sociologie*, 34(4), 495–524.

Larner, W. 2003. Neoliberalism? *Environment and Planning D*, 21(5), 509–12.

Larner, W. and Walters, W. 2004. Global governmentality: governing international spaces, in *Global governmentality*, edited by W. Larner and W. Walters. Milton Park: Routledge, 1–20.

Lohmann, L. 2005. Marketing and making carbon dumps: Commodification, calculation and counterfactuals in climate change mitigation. *Science as Culture*, 14(3), 203–35.

Lohmann, L. 2010. Neoliberalism and the calculable world: The rise of carbon trading, in *The Rise and Fall of Neoliberalism: The Collapse of an Economic Order?*, edited by K. Birch and V. Mykhnenko. London: Zed Books, 77–93.

MacKenzie, D. 2009a. Constructing emissions markets, in *Material Markets: How Economic Agents Are Constructed*. Oxford: Oxford University Press, 137–76.

MacKenzie, D. 2009b. 'Making things the same: Gasses, emission rights and the politics of carbon markets'. *Accounting, Organization and Society* 34(3–4): 440–55.

McCarthy, J. and Prudham, S. 2004. Neoliberal nature and the nature of neoliberalism. *Geoforum*, 35(3), 275–83.

Marres, N. 2008. The making of climate publics: Eco-homes as material devices of publicity. *Distinktion: Scandinavian Journal of Social Theory*, 9(1), 27–45.

Meyer, J., Frank, D., Hironaka, A., Schofer, E. and Tuma, N. 1997. The structuring of a world environmental regime, 1870–1990. *International Organization*, 51(4), 623–51.

Moore, K. Kleinman, D., Hess, D. and Frickel, S. 2011. Science and neoliberal globalization: a political sociological approach. *Theory and Society*, 40(5), 505–32.

Oels, A. 2005. Rendering climate change governable: From biopower to advanced liberal government? *Journal of Environmental Policy and Planning*, 7(3), 185–207.

Ong, A. 2006. *Neoliberalism as Exception: Mutations in Citizenship and Sovereignty*. Durham, NC: Duke University Press.

Osborne, T. and Rose, N. 1999. Do the social sciences create phenomena? The example of public opinion research. *British Journal of Sociology*, 50(3), 367–96.

Peck, J. and Tickell, A. 2002. Neoliberalizing space. *Antipode*, 34(3): 380–404.

Rose, N. and Miller, P. 1992. Political power beyond the state: Problematics of government. *British Journal of Sociology*, 43(2), 173–205.

Rutherford, S. 2007. Green governmentality: Insights and opportunities in the study of nature's rule. *Progress in Human Geography*, 31(3), 291–307.

Sethi, N. 2007. Poor joke on carbon credits: NGO transfers Indian farmers' green points to British globetrotters. *The Times of India*, 20 September.

Swyngedouw, E. 2010. Apocalypse forever? Post-political populism and the spectre of climate change. *Theory, Culture and Society*, 27(2–3), 213–32.

Thévenot, L., Moody, M. and Lafaye, C. 2000. Forms of valuing nature: Arguments and modes of justification in French and American environmental disputes, in *Rethinking Comparative Cultural Sociology: Repertoires of Evaluation in France and the United States*, edited by M. Lamont and L. Thévenot. Cambridge: Cambridge University Press, 229–72.

Thévenot, L. 2002. Which road to follow? The moral complexity of an 'equipped' humanity, in *Complexities: Social Studies of Knowledge Practices*, edited by J. Law and A. Mol. Durham, NC: Duke University Press, 53–87.

Tsing, A. 2005. *Friction. An Ethnography of Global Connection*. Princeton, NJ: Princeton University Press.

Voss, J.-P. 2007. Innovation processes in governance: The development of 'emission trading' as a new policy instrument. *Science and Public Policy*, 34(5), 329–43.

Wagner, P. 2008. *Modernity as Experience and Interpretation. A New Sociology of Modernity*. Cambridge: Polity Press.

World Bank. 2010. *State and Trends of the Carbon Market 2010*. Washington, DC: Carbon Finance.

Chapter 9

The green transition, neoliberalism, and the technosciences

David J. Hess

Introduction

A political sociological approach to the problem of neoliberalism and the technosciences builds on but is distinct from two other influential frameworks in the social sciences: cultural analysis, which studies practices and discourses such as entrepreneurialism across social fields and over time (e.g. Foucault 2008), and class analysis, which studies the shifts in government policy since the 1970s as a strategy advanced by large capital (e.g. Harvey 2005). Rather than focusing primarily on cultural practices or class conflict, political sociology situates the problem of neoliberalism and technoscience within quasi-autonomous but interrelated social fields. Our previous work on the topic identified three primary fields: the university, which has undergone a transformation to academic capitalism and an asymmetric convergence with the private sector; civil society, which has politicized the research agendas of the sciences, posed alternatives to the design directions of technologies, and in general modernized the relationship between researchers and their publics; and the state, for which regulatory policy has undergone a process of scientization, privatization, and internationalization (Moore et al. 2011). This chapter develops the line of research in the third field, the state; within that topic the focus is on the problem of trade and industrial policy rather than regulatory policy, and within trade and industrial policy the attention is on the regulation of green technology.

The argument about the relationship between neoliberalism and the technosciences is developed in three parts. First, the political field is described as a contested social space in which agents draw on underlying political ideologies that shape and are shaped by their action (Bourdieu 2005, Fligstein and McAdam 2011). Thus, neoliberalism is one of other contending ideologies rather than a totalizing policy regime. Second, an analysis of the green-energy policy field in the United States is used to show how the ideologies are embedded in specific policy directions and decisions. Third, the problem of how to relate one type of green technology, solar energy, to the underlying ideologies of the political field is discussed as a way of formulating the problem of the relationship between neoliberalism and the technosciences. The analysis will be limited to the United States and the case of green technology, but the case study will be used to develop the general argument that there is a historically contingent relationship among a research program in the scientific field, a type of technology that appears in an industrial field, and a political ideology in the political field. Steering a path between technological determinism and unlimited interpretive flexibility, I suggest the value of a perspective that attends to broad issues of differences in sociotechnical systems design and their homologies with positions in the political field.

The analysis presented here is based on previous work on localist movements and green-energy policies in the United States. The research projects were based on summer training seminars for graduate students and interviews by the author, with a total of about 80 interviews and case studies of green-energy policy in the American states and 20 cities. Interviews were supplemented by studies of government documents and attendance at approximately two-dozen conferences.

Neoliberalism in the political field

Although one might define neoliberalism narrowly as a subfield of economics associated with monetarist and supply-side approaches to economic policy, the approach taken here is to view neoliberalism more broadly as a political ideology, that is, as a network of cultural models of and for action that provides a point of reference to guide and legitimate action in the political field. Tax-cutting, anti-regulatory, and anti-inflationary policies have a place in this conceptualization of neoliberalism, but it is also understood to involve a broader political philosophy associated with the principles of decision-making by markets, limited government, fiscal discipline, privatization, devolution, and the importance of self-responsibility. As an ideology, neoliberalism is situated in a contested political field that includes competing ideologies, but the ideologies often are not manifest in their pure form. Instead, the view of a particular political agent or embedded in a specific policy often is a compromise formation that bears the imprint of more than one ideology. Thus, for the purposes of social scientific analysis, ideologies are ideal types that facilitate an analysis of the relations among positions in the political field.

One of the great problems with much current scholarship on neoliberalism is that it does not pose limitations on the concept. At its worst, the scholarship can result in a totalizing analysis in which almost all historical changes that have occurred since the 1970s are part of a neoliberalization process. In contrast, a field sociological approach begins with the assumption that neoliberalism is one among other positions in the political field. However, just as one can analyse the position of agents in a political field as having a relatively dominant or subordinate position, so one can analyse transitions in the political field in which one ideology becomes hegemonic. This section will provide a map of the ideologies in the United States and their relative position in the political field.

To begin, neoliberalism is only one type of liberalism, that is, an ideology based on the premise that there should be limits on the role of the state in the economy, which is assumed to be composed largely of private firms related to each other via markets. Historically, the concept of liberalism is associated with the limitations on the rights of monarchs and nobles that emerged in the constitutional developments in early modern Europe. Within the long tradition of liberal thought and political practice, important tensions emerged during the late nineteenth and early twentieth centuries, when liberal political orders responded to the challenge of socialism. Thus, the primary alternative to liberalism in the modern political field is not royal absolutism but socialism, or the government control over the economy through state ownership of enterprises. Although other alternatives exist, such as the anarchist organization of society into communes, only socialism has played a major role in world history during the twentieth century. Although considered to be defeated and discredited after the conversion of communist countries to capitalist and quasi-capitalist economies, it is also the case that public ownership of some industries continues to play a significant role in many of the leading national economies. In the United States, public ownership is highly restricted (for example nuclear energy, roads, passenger intercity rail, and the postal service), whereas in other countries public ownership often includes the health-care system and mineral extraction industries.

Within the mainstream of twentieth-century liberalism, the primary tension was between neoliberalism and what can be called a social liberal (or, in Europe, social democratic) variant of liberalism. In the United States social liberalism is associated with the policies of the New Deal, Great Society, and President Nixon. The presidency of Jimmy Carter included many reforms that, in retrospect, suggested a shift in political currents in favour of neoliberalism, and the presidency of Ronald Reagan marked the first full expression of neoliberalism in national politics. When the Democratic Party returned to power in the 1990s, there were strong currents of neoliberalism in many policies, such as the transformation of the welfare system.

The central tension between social liberalism and neoliberalism involves the relative role of the state in the economy. The ideal-typical social liberal believes in using the state to provide some economic redistribution, some public ownership, regulatory intervention for market failure, protections for labour and the poor, and deficit spending in periods of an economic downturn. Although social liberalism is closely associated with Keynesianism in economic policy, it is a broader ideology that goes beyond the idea that government intervention is needed to smooth out the economic cycle and move the economy to a desirable general equilibrium. In contrast, the ideal-typical neoliberal believes in a much more restricted role for the government and generally advocates reversing social liberal policies in favour of decision-making through market mechanisms. Economic policy is focused on reducing government spending, often by cutting redistributive programs and taxes and by reducing regulatory burdens on the private sector. The supply-side approach to economic policy is intended to unleash the governmental fetters on the private sector and enable growth, and likewise the neoliberal focus on price stability enables a healthy, long-term climate for investment decisions. In some cases there is also support for government intervention in the economy in order to create new markets and competition, often in place of highly regulated policy regimes (such as for airlines and electricity in the United States).

Within the United States, social liberalism and neoliberalism represent two poles of the mainstream of political debate. One tends to find political leaders that come close to the ideal type of social liberalism on the left wing of the Democratic Party, such as politicians associated with labour and progressive coalitions, and one tends to find leaders who approximate the ideal type of neoliberalism on the right wing of the Republican Party, such as Tea Party activists. Often leaders in the political centre from both parties develop compromise policies that can be analysed as having social liberal and neoliberal components. Over time the political centre has shifted to the right, and leaders of the Democratic Party have increasingly adopted positions that are consistent with neoliberalism.

On the margins of the political field, in subordinate positions in most policy fields, one can find radical ideologies such as socialism and localism. Although social liberals accept some public ownership, socialists advocate for a much higher level of public ownership, for example in health care, transportation, and energy. Currently, there is only one openly socialist senator in Congress, and there has not been a significant socialist presidential candidate since the Citizens Party of Barry Commoner in 1980. Localism is a more ascendant political ideology; it is associated with mobilizations by the small-business sector to achieve access to capital and protections from predation by large capital such as chain stores. Localist politics can be traced back to the coalitions between farmers and small businesses that characterized the populist and progressive movements of the late nineteenth and early twentieth centuries. Those coalitions were often linked to the trade-union movement, and they were swept out of the mainstream of American politics by the New Deal coalition, which developed a social compact between the industrial unions and large capital. Localism in the twenty-first century can take the form of locally owned, independent retailers who are mobilized in support of the 'buy local' movement and anti-big-box mobilizations, but it can also take a more progressive, 'local living economy' vision of small businesses that address issues of social fairness and sustainability (Hess 2009). Support for the local living-economy vision is found among independent, green retail stores (such as fair trade coffee shops); food cooperatives; small farms that are linked to local consumers through independent restaurants and farmers' markets; independent media; credit unions and community banks; reuse centres; advocates of human-powered and public transportation; and systems of distributed and locally controlled renewable energy.

Developmentalism

The typology of ideologies in the political field is adequate up to a point; however, I have found that in the analysis of green-energy policies in the United States it has been necessary to characterize another position: developmentalism. Unlike localism and socialism, the field position is often in the political mainstream, especially at the state-government level. Developmentalism in the United States is liberal in that it retains a vision of the economy based on firms, and it is mainstream in the sense that it assumes an economy characterized by corporate capital rather than small businesses or public enterprises. However, in contrast with the two other mainstream ideologies in the political field, social liberalism and neoliberalism, developmentalism involves a much more defensive approach to free trade and further trade liberalization. Furthermore, redistributive politics are approached neither through welfare-state programs nor through supply-side entrepreneurialism but instead through job-creating industrial policy. Although not isolationist, the politics of developmentalism include emergent protectionism and a call for government intervention in the private sector to enable competitiveness. Those two sides indicate a Janus quality to developmentalism: it seeks both protections from a liberalized global economy and enhanced competitiveness to enable more effective participation in it.

The rise of developmentalism in American politics is due to the effects of the long-term pattern of trade liberalization on the American economy (Hess 2012). Since World War II, there has been an underlying agreement between social liberals and neoliberals that trade liberalization is beneficial to the American economy and to the rest of the world, and the post-war economic order set in motion a mechanism for ongoing trade negotiations with the goal of attaining higher levels of trade liberalization. No matter how interventionist Keynesians, New Deal Democrats, and other social liberals were with respect to domestic policies, they believed in free trade, and they assumed that protectionist policies had worsened the trajectory of the Great Depression and enhanced the possibility of war. However, during the period of the hegemony of social liberals in national politics (1932 to 1980 in the United States), there was tolerance for protectionist policies among less developed countries. According to economist Ha-Joon Chang (2011), during the 1950s and 1960s trade agreements focused on developed countries, and less developed countries were allowed to support infant industries by enacting developmentalist policies such as import substitution, tariff barriers, currency devaluation, government support and ownership, and capital controls. After the economic stagnation and the mobilization of Third World countries during the 1970s, wealthy countries extended the scope of trade liberalization to include formerly protected, poorer countries. Where those countries could not be convinced of the benefits of trade liberalization, they were often forced to accept such conditions as part of structural adjustment packages negotiated for debt repayment.

Within both wealthy and newly industrializing countries, trade liberalization facilitated the underlying economic conditions that transformed the political field in a direction that generally favoured advocates of neoliberal policies. As large capital became globalized, the threat of run-away shops and competition from foreign enterprises increased the bargaining power of capital with respect to the state and labour. Large capital was then able to make successful demands on government to reduce the regulations and tax policies associated with the social liberal compact of the Great Depression, and likewise it was able to wrest from labour reductions wage and other demands from capital. In turn, the loss of revenues from lower taxes, privatization, and lower wages placed financial pressure on the state. The old social compacts of social liberalism could be temporarily preserved through deficit spending, but in the long term the resolution of the revenue gaps occurred by reducing expenditures associated with the redistributive functions of the welfare state and by selling off public assets. The whittling away of the social liberal state occurred both incrementally over decades and suddenly during fiscal crises, and it occurred with different compromise formations with the old social compact in different countries (Harvey 2005). Supporters of social liberal ideology never disappeared, and specific laws and policy reforms after the rise of neoliberal ideology to a relatively dominant position in the political field often emerged as compromise formations that showed evidence of both social liberal and neoliberal ideologies. Nevertheless, in many countries there was an historic change in which political actors associated with neoliberal ideology often achieved a dominant position in national governments and in specific policy fields.

My argument with respect to developmentalism is that just as trade liberalization weakened the political position of social liberals, it is now undermining the position of neoliberals and opening the political opportunity structure to the ascendency of developmentalist liberalism. Neoliberals' advocacy of opening up the global economy to trade with less developed countries has led to the rapid growth of some of the countries, above all China, which in turn has led to the relative decline of the United States as the hegemonic economy in the world system. China has increased its position in the global economy through developmentalist policies, including currency devaluation, import restrictions, and subsidies for export-oriented businesses. Recognition of the aggressive use of developmentalist policies by the country's primary global rival has strengthened the growing chorus of voices that call for a more developmentalist approach to American economic policy. The idea that the United States could compensate for the loss of low-tech manufacturing jobs to newly industrializing countries by climbing up the ladder of technological complexity to high-tech jobs (and to protect the high-tech sector through trade-related intellectual property agreements) has proven to be elusive, because China has rapidly moved into high-tech manufacturing as well.

Although the liberalization of trade has enabled economic elites in the United States to benefit from participation in the high-technology export market and from investments in foreign countries, the changes in the American economy have left behind the working class and increasingly also the lower professional strata. As economists Michael Spence and Sandile Hlatshwayo (2011) argue, wages have increased in the tradeable or export-oriented sector, but job growth has been limited. In contrast, in the nontradeable or domestically oriented sector, there has been job growth but wage compression. More generally, for hourly workers, there has been wage stagnation for decades, and overall inequality in the country has increased (Labor Research Organization 2004). The declining position of working people has in turn generated a growing trend of scepticism toward free trade, which is seen as taking away good jobs. Public opinion polls in the United States show increasing lack of faith in free trade and increasing concern that it has taken away jobs. The percentage of Americans who believe that free trade had hurt the country from 30% in 1999 to 53% in 2010, and 69% believed that free-trade costs Americans jobs. Opposition to free trade was also high among both Tea Party and union supporters, a pattern that suggested a place where right-left polarities were being cross-cut by a different ideological current (Harwood 2010).

Scholars committed to the historical thesis of ongoing and deepening neoliberalization would likely argue that the resurgence of developmentalism is a further deepening of neoliberalization, and in contrast those who are committed to the historical thesis that neoliberalism is in a state of crisis would lean toward an interpretation of developmentalism as a fundamentally different form of liberalism. From a field sociological perspective, developmentalist politics and policies are an analytically distinct type, due primarily to fundamental differences with respect to the underlying doxa of support for free trade in both social liberalism and neoliberalism. In practice developmentalist approaches may be aligned with further neoliberalization, the oppositional politics of social liberalism, or even marginal positions in the political field (such as localism, which has a strong current of advocacy for local import-substitution). Thus, from an analytical perspective it is best to treat developmentalism as a separate ideological current in the political field rather than merely a development within neoliberalism.

Political ideology and the green-energy transition

As the global economy in the twenty-first century adjusts to the relative decline of the liberal triad of the twentieth century (Europe, North America, and Japan) and the emergent defensiveness in the United States toward trade, it also faces a second, interlocking historical transition: adjusting the material infrastructure of the global economy to the global ecology. There is general recognition that the combination of human population growth and high consumption levels has exceeded the planet's carrying capacity. More than global warming, there are interlocking ecological crises that include deforestation, desertification, peak oil, freshwater depletion, and marine life extinction.

Naomi Klein (2011) has argued that climate science and policy (and, by extension, all sustainability sciences) are at some level anathema to the politics of neoliberalism, because environmental and energy problems are so extensive and pervasive that they require government planning and intergovernmental coordination to solve. To some degree the valences toward government intervention and steering in the economy help explain the growing anti-green reaction in the United States, which has been mounted under an anti-government, pro-market ideology associated with Republican Tea Party activists. Although in part a grassroots movement, wealthy benefactors from the oil and coal industries, which have the most to lose from a rapid green transition, have supported Tea Party organizations and political candidates and provided legislators with policy guidelines. In the United States, the coalition of conservative organizations and fossil-fuel industries has created a division within large capital between the anti-green industries and other industries that have embraced the idea of a long-term green transition as part of corporate strategy. For example, General Electric, venture capital from Silicon Valley, and some of the large information-technology and consumer-products firms have become supporters of stronger green-transition policies, partly because they have invested significantly in green technology. During the 2010 election the California venture capitalists, in alignment with labour unions and environmental organizations, played a significant role in turning back a challenge to California's green-transition policies that had been funded by out-of-state fossil-fuel interests. Even when the national political field turned anti-green in 2010 under the wave of climate-denying political leaders, green-transition policies continued to deepen at the state and local government level where the Democratic Party remained firmly in power (Hess 2012).

Although I agree with Klein that green-energy policies (as part of a broader transition to a more sustainable economy) are caught in the intense ideological crossfire of the political field, I do not agree that green-energy policies as a whole are inherently opposed to neoliberalism. Rather, I suggest that green-energy policies themselves are transected by the different ideologies in the political field. Thus, the categorization of the political field in the previous section can be used to interpret the politics of green-energy policies.

Social liberalism. A portion of the green-energy policy field is closely associated with the redistributive politics of social liberalism. The most prominent example is the creation of green-jobs training programs that are oriented toward persons with employment barriers, including at-risk youth. Programs can be found in city and state green-jobs corps and nonprofit organizations that promote such policies, such as Green For All. The programs provide training so that the unemployed can find positions in grounds maintenance, weatherization of buildings, rooftop solar installations, and other green service-sector industries (Hess et al. 2010). Government subsidies to renters and owners of apartment buildings enable them to retrofit buildings in ways that reduce the burden of energy costs. Although some of the policies are in the New Deal tradition of direct government employment, many of the programs represent a modified form of social liberalism that channels government funding into nonprofit training centres, contracts for private businesses, and public-private partnerships. In some cases labour unions receive funds for training, and workers may become members of unions such as the Laborers' International Union of North America. Thus, this variant of green-energy policies speaks directly to labour-environmental-urban poverty coalitions that are the traditional constituency of social liberal politics.

Socialism. Public ownership in the energy field is the legacy of two previous eras of energy policy. During the height of the progressive era, some city governments municipalized their electricity generation and distribution system, and some publicly owned municipal electricity systems, such as Seattle City Light, are a legacy of the period. During the height of social liberalism, state governments such as New York and the federal government also created publicly-owned electricity generation systems, at first mostly for hydropower and later for nuclear energy. Although most of the electricity in the United States is produced through publicly regulated, private enterprises called investor-owned utilities, about forty-six million people are served by public power (American Public Power Association 2011).

Some of the greenest energy distribution and generation organizations in the country are publicly owned municipal organizations. For example, Seattle City Light became the country's first carbon-neutral electricity service provider based on generation from solar, wind, and hydropower in combination with purchased carbon offsets. The organization also utilizes net revenue to invest in building retrofits and distributed renewable energy such as rooftop solar. Studies of public power indicate that they generally have better performance on greening metrics in comparison with investor-owned utilities (Hess 2009).

Localism. There is a tradition of cooperative ownership of electricity generation in the United States, generally in rural areas, and some city governments and municipal electricity service providers also offer ratepayers an opportunity to buy shares in a solar array, which enables people to have the benefits of local ownership in solar-energy generation without the problems of rooftop maintenance. However, by far the most widely diffused policy associated with local ownership was property-assessed clean-energy (PACE) financing, which enabled owners of homes and commercial buildings to pay for rooftop solar and building retrofits with an increment on their tax bill that in turn was financed by a city or state revenue bond. The program offered the advantages of affordable financing and liquidity in the event that the owner must sell the building. Although in 2010 the federal housing authorities ended PACE programs because the property tax lien was potentially in conflict with their mortgage lien priority, the programs were continued for commercial buildings and in some cases for residential buildings but with secondary lien status. Furthermore, a similar program developed in New York allows for on-bill financing through the utility company (Van Nostrand 2011).

Although social liberalism, socialism, and localism play a role in green-energy politics and policies, there is also evidence that developmentalist and neoliberal ideologies have played a major role. Their role is discussed in the next two sections.

Developmentalism and green-energy politics

A primary example of the protectionist dimension of developmentalism during the Obama administration is the 'Buy America' provision in the American Recovery and Reinvestment Act of 2009. The provision was designed to favour American companies in infrastructure projects, some of which involved green-energy construction such as wind farms. The policies were designed in way that did not antagonize American trading partners that had signed the Government Procurement Agreement, a side agreement of the World Trade Organization that ends domestic content provisions for federal government purchases. Because China and some other newly industrializing countries had not signed the agreement, the Buy America provision cleverly served as a protectionist measure directed toward those countries, but one that was allowed under international law. Another example of the defensive posture to trade that is characteristic of developmentalism is the complaint that the United States filed with the World Trade Organization to pursue China's allegedly illegal support of green-technology companies in the export sector (Chan and Bradsher 2010). The United States and China have also invoked trade sanctions on many specific product categories, including green-energy products. Finally, American support for some green-technology companies, such as battery and wind-turbine manufacturers, has allowed the country to regain some manufacturing that had been lost to Asia. The support is legal under World Trade Organization rules provided that the companies are producing for domestic markets. General support, such as research funding through federal agencies and the national laboratories, is also allowed.

In 2011 the American solar-energy industry became the main arena in which the politics of developmentalism were debated in Congress. The bankruptcy of Solyndra served as an occasion for neoliberals to decry the Obama administration's green developmentalism as a failed industrial policy. For example, Republicans attacked industrial policy as a role for government best left to markets, and they questioned the propriety of using taxpayer funds to play the role of venture capitalist (Wald and Savage 2011). The president defended the failed investment as part of the necessary risks that a government must take, and he argued that it was important for the United States to maintain a solar industry. The solar industry also argued that China was attempting to take over the world's solar industry through subsidies that are not legal under WTO rules. As Gordon Brinser, president of SolarWorld Industries America, Inc., stated, 'The illegal subsidies that China has put into the solar industry are no different than giving an athlete a bucket full of steroids' (Drajem and Martin 2011). The solar industry also asked the American government to put duties of more than $1 billion on Chinese solar imports. The leading Chinese solar manufacturer, Suntech, responded by claiming that the American solar manufacturers were acting in a 'protectionist' manner that could trigger a 'solar trade war' that would harm the country's goal of 'achieving a clean-energy future' (Drajem and Martin 2011).

This type of exchange between the United States and China has become commonplace. The Chinese argue that their successes are due to consistent support over time in contrast with the inconsistences in American policy across political administrations. They also note that the American government is hypocritical because it also has 'buy American' provisions. In response, Americans argue that Chinese policies are far more onerous and in violation of trade agreements. For example, Leo Hindery, chairman of the US Economy/Smart Globalization Initiative of the New American Foundation, explained how China's Indigenous Innovation Production Accreditation (IIPA) Program affected American companies. He noted that the program is 'far more restrictive than any other buy-domestic program in the world', because it requires companies that want to sell their products in China to 'transfer and license, to Chinese companies, their latest technology for "co-innovation" and "re-innovation"' (Hindery 2010). Hindery added that foreign companies in China have also been unwilling to complain, because they are afraid of retaliation from the Chinese government.

Although the solar industry has become central in the debate over developmentalist policies, evidence of developmentalism can be found in other energy policy fields as well, such as liquid fuels. Federal government policies that support the transition away from petroleum, such as support for electric vehicles, are based on a strategy of import-substituting industrialization that is usually framed under the banner of energy security. Likewise, the federal renewable fuels standard, which is combined with an ethanol tariff, supports the development of the domestic ethanol industry, even over more efficient foreign alternatives, such as the Brazilian ethanol industry. Support for American biofuels is justified by the infant-industry argument that eventually the American industry will transition to a more efficient, cellulosic form of ethanol and algae-based biodiesel. In 2010 the Brazilian government removed its 20% tariff on ethanol as part of an effort to get the American government to open up its protected ethanol industry to free trade. In turn, the American biofuels lobby rallied to keep its protections in place. Industry spokesperson General Wesley Clark noted, 'Ethanol is America's fuel: It's made here in the U.S., it creates U.S. jobs, and it contributes to America's national and economic security' (Mulkern 2010).

In battles over the American solar and biofuels industries, there is an interesting role-reversal on the issue of free trade. The world's historical home of free-trade rhetoric and policy, the United States, has found itself defending its industries against the calls for more openness from emerging industrial giants China and Brazil. In turn, those countries have aggressively pursued industrial and trade policies to support their domestic industries. As threats to American industries mount, the country is pushed into a situation of having to defend those industries through trade protection and industrial policy support. The situation is a long-term change, and one can find evidence of developmentalism even during the 1980s, when the American government responded to protect its automotive industry from Japanese competition. However, the level and extent of global industrial competition has continued to rise, and with it so has the motivation for developmentalist policies.

Defensive trade policies and import-substitution policies for petroleum tend to receive bipartisan support in the federal government, because they do not provoke sectional rivalries and they enable a bipartisan consensus to be built in opposition to perceived foreign threats. In contrast, industrial policies such as support for solar energy tend to be more controversial, partly because they flounder on sectional rivalries and partly because they involve a heavy government hand in the steering of the economy, which neoliberals oppose on philosophical grounds (Eisinger 1986). As a result, the industrial policy side of developmentalism in the United States tends to be more common at the state-government level.

State governments aggressively pursue green developmentalism both through import-substitution policies and green industrial policy. State governments have engaged in an import-substitution policy for energy by replacing coal and natural gas that are 'imported' from other states with local renewable energy and, in the case of oil from other countries, with local biofuels and electricity. To support the transition, state governments have developed renewable-electricity portfolio standards, demand support from system benefits charges, and carbon trading in the Northeast and California. Likewise, energy-efficiency measures, such as the numerous building-efficiency laws and ordinances passed at the state and local government level, replace imported energy with local labour and sometimes locally manufactured building materials. Although some of the policies were first approved based on environmental motivations, increasingly they have been reframed and justified on economic-development and job-creation grounds. In addition to the more implicit policies of import substitution, some state governments have also enacted measures that support in-state manufacturing and production. For example, some states now have a goal that biofuel consumption in their state will include a percentage produced from in-state biofuel sources. Florida has also pioneered biofuel production from citrus peels, a source that is relatively unique to that state. We also identified state governments that have created incentives for wind and solar production projects that motivate the use of wind and solar components that are manufactured in-state (Hess et al. 2010).

The other side of developmentalism, industrial policy, is especially visible among state and local governments. Rather than attract a large manufacturer that might later decide to pull up stakes and move to another region in the world, economic development strategies focus on growing local businesses by creating dense networks of government support, university-based research, technology transfer, business services, venture capital, and jobs training programs. The goal of the cluster is to make the region itself a reason for businesses to stay in place. Even if they eventually locate some of their manufacturing operations elsewhere in the world, they are likely to retain high-end functions such as management, research, and marketing in the region. The clusters are based on local industrial histories and factor endowments, and increasingly they are the outcome of government planning that targets specific regional industries as candidates for regional industrial policy. Green-tech industries have begun to concentrate in some regions of the country through regional industrial clusters for wind (Colorado, Iowa, Michigan, South Carolina), fuel cells (Connecticut, New York, Ohio), solar (California, Ohio), battery and energy storage (Michigan, New York), and rail (New York) (Hess et al. 2010).

Although the clusters are poised to participate in the global economy, they also benefit from local demand policies, such as solar and wind carve-outs in the renewable electricity standards. Cluster development also requires a suite of supply-side policies, such as the cultivation of local networks, finding finance for start-ups and business growth, providing green jobs training (at a higher skill level than the programs discussed above), and connecting local research universities with the business community. These highly interventionist policies are anathema to neoliberals, who reject the 'horse betting' of industrial policy, but they are also distinct from the redistributive policies associated with social liberalism. Nevertheless, because the policies promise jobs, and often good jobs in high-tech manufacturing, they receive support from segments of unionized labour and corporate capital.

Neoliberalism and green-energy politics

Within this broad field of policy reform, neoliberal politics have played both a generative role in promoting green-energy policies and an oppositional role to their development. With respect to the generative role, three types of electricity policies associated with neoliberalism have opened up opportunities for renewable energy. First, government reforms have led to the slow restructuring of the electricity industry, beginning with the Public Utility Regulatory Policy Act in 1978. The liberalization of the industry during the 1990s created new markets that enabled competition among electricity generation companies and retail service providers. The changes created some political opportunities for green-energy production, because some state governments responded to the new regulatory environment by requiring utilities to purchase green electricity from independent generators or produce it themselves.

A second approach that is consistent with neoliberal ideology has been the removal of barriers to distributed energy production. Regulatory changes such as net metering enabled the interconnection between energy producers and the grid, and regulations that reduced the burdens of permitting (called 'fair permitting' laws) have facilitated the greening of electricity. Generally, the laws also protected producers by mandating that utilities purchase power at the avoided cost rate, a more market-oriented approach than the feed-in tariff, which is more common in Europe. Our review of party-line votes in state legislatures also indicates that this approach to green-energy policy tends to receive bipartisan support (Hess 2012).

A third approach has been to implement environmental policies by creating new markets. Thus, the general pattern for carbon regulation in the United States has been based on a cap-and-trade system rather than a carbon tax. Likewise, some of the states that have a renewable-energy portfolio standard have experimented with markets for renewable energy credits.

Although neoliberal ideology has played a generative role in green-energy transition policies, it has also inspired opposition to them. The hostility is largely found among anti-green political candidates who have been supported by the coalition of conservative and fossil-fuel donors. Although conservative foundations had long funded climate-change scepticism, the mobilization after the election of President Obama and the attempt by Democrats to pass a national cap-and-trade law was of a different order, because it involved attempts to roll-back existing green-energy policies.

With respect to social liberal, localist, and social liberal policies, the attacks by neoliberals have been relatively muted. The green-jobs training programs for persons with employment barriers represent relatively small expenditures, and they also represent third-way social liberalism, in which traditional welfare and antipoverty programs have been combined with the more neoliberal philosophy of helping the unemployed to find work. Based on our review of green-energy legislative votes in state governments, the localist programs such as PACE and on-bill financing have also received bipartisan support in state legislatures, perhaps because they do not involve increased taxes or fees, and because they are ultimately based on a voluntary choice that creates opportunities for business development (Coley and Hess 2012). In contrast, recent attempts to municipalize local energy, such as the long battle in San Francisco that emerged in response to the crisis in 2000 and 2001 caused by electricity restructuring, have met with strong opposition from utilities, which have attacked such efforts as socialistic (Hess 2009).

The more concentrated attacks that are anchored in a neoliberal ideology have been reserved for developmentalist policies. Attacking demand policies such as cap-and-trade programs, renewable electricity standards, and system benefits charges, Tea Party activists and other Republican leaders argued that they were taxes that placed unnecessary burdens on businesses and households. In 2011, the state of New Jersey withdrew from the regional greenhouse gas initiative, the state of New Hampshire remained in the agreement only after a razor-thin vote, and the effort among Midwestern states to develop similar initiatives stalled. As noted above, in 2011 anti-green Republicans took advantage of the collapse of the solar-energy manufacturer Solyndra to attack green industrial policy in general. Thus, whereas scholars of neoliberalism have viewed 'roll-back' strategies as a phase of neoliberalization associated with the 1980s (e.g. Peck and Tickell 2002), one can use the case of the anti-green backlash to build on their work by noting that roll-back neoliberalism has received a second life. The attacks on developmentalist policies both on the demand side (unnecessary taxes) and the supply side (wasteful and misconceived industrial policy) suggest the need to view developmentalism as distinct from neoliberalism, at least in the case of the United States during this historical period.

One can also see the debate in general ideological terms among some of the political commentators. For example, Nobel laureate and *New York Times* columnist Paul Krugman has defended a developmentalist approach to American industry, including protectionism:

> There's the claim that protectionism is always a bad thing, in any circumstances. If that's what you believe, however, you learned Econ 101 from the wrong people — because when unemployment is high and the government can't restore full employment, the usual rules don't apply.

> Let me quote from a classic paper by the late Paul Samuelson, who more or less created modern economics: 'With employment less than full [...] all the debunked mercantilistic arguments' — that is, claims that nations who subsidize their exports effectively steal jobs from other countries — 'turn out to be valid'. (Krugman 2009)

In contrast, Robert Samuelson, a contributing editor for the *Washington Post* and *Newsweek* (and not related to Paul Samuelson), worries about the decline in faith in free trade in the United States:

> Of course, opposition is not new. Even if free trade benefits most countries, some firms and workers lose from added competition. But for most of the postwar era, a pro-trade consensus neutralized this opposition. That consensus is now fraying. (Samuelson 2007)

Science, technology, and neoliberalism

With the map of ideological positions in the political field available as a point of reference, it is now possible to turn to the problem of how neoliberalism is related to science and technology. Distinctions within economics research—for example, among Keynesian, supply-side, and heterodox schools associated with studies of local ownership, public ownership, and import substitution—can be easily linked to positions in the political field. Likewise, it is relatively easy to show general connections between the overall valuation of research fields and political ideology. As Bourdieu (1998) argued, the right and left hand of the state are the congealed outcome of many historical battles, including the battles of social movements to achieve limitations on the power of global capital over both small businesses and social life in general. To extend Bourdieu's argument, the scientific field also has a right and left hand. In the case of university-based research there are schools and departments in relatively close alignment with the needs and interests of the corporate clients and benefactors (business, medicine, engineering, applied sciences), and there are schools and departments more aligned with the left hand of social liberalism (social work, sociology, geography, anthropology, public health, some humanities, and schools and departments associated with community development). As the state withdraws from its left-handed functions, the university also shrinks its left hand, resulting in some cases in the dismemberment of traditional left-handed departments. In a process that mirrors the asset stripping of governments, universities in the United States have in some cases closed departments and programs in the humanities and social sciences, even as they have invested in schools associated with high technology. The case of the State University at Albany is a good example: the university and state government have invested heavily in a new school of nanosciences, but the university also closed several departments in the humanities.

Can one extend the historical restructuring of the priorities of university research fields to make a general argument that the technosciences that have emerged since the 1980s and that have been the benefactors of academic capitalism—nanotechnology, information technology, biotechnology, and green technology—are in some ways inflected by neoliberalism? A field sociological approach to the question would suggest caution against blanket statements (for example biotechnology research is inflected as a whole by neoliberal politics), and the case of green technology can be used to demonstrate a method for studying the relationship that is more disaggregated. In this section I will take the example of solar energy and examine it as a sociotechnical system, that is, as a system that includes technical design choices and organizational, user, and regulatory choices (Hughes 1983, 1987).

Beginning first with the choice between a large-scale, grid-based solar farm and a small-scale, building-based rooftop system, how do the design differences align with ideological positions? The social liberal approach to green-energy development would have a valence in favour of the rooftop configuration, which would favour the green-jobs programs linked to retrofitting programs, where the politics of social liberalism have most closely connected with green-energy policies. Likewise, localism would also tend to favour the rooftop scale, because it is consistent with local ownership. Municipal utilities might favour the centralized solar farm, because it would easier and less costly to develop and manage than a distributed system of similar scale. Supporters of developmentalism are potentially neutral on the topic. Large contracts are easier to align with manufacturing needs, but distributed energy is aligned with state government goals to substitute renewable energy for out-of-state, fossil-fuel sources. Finally, a neoliberal approach may favour distributed renewable energy, because it is more consistent with a voluntary, market-oriented approach, whereas solar farms tend to be constructed only in the context of government intervention in electricity 'markets' in the form of renewable electricity standards, system benefits charges, and carbon-based regulatory caps.

In short, on the issue of the scale of solar technology, there are valences in the design decision that would enable an analyst to develop what Winner (1986) called a politics of design. This strategy of analysis goes beyond the more general statement that would argue that solar-energy technology as a whole has a single political valence. Thus, one avoids the blanket statement that 'green-tech' is in some general way inflected by neoliberalism. Instead, one looks at positions within the field of a specific green technology and relates it to positions in the policy field.

One can complicate the analysis further by studying design distinctions at one scale. For example, at the scale of grid-based solar farms, there are also design choices between the concentrating solar (steam-powered) approach and photovoltaic farms. The former is a very large and heavy technology that may be more amenable to domestic manufacturing (like the larger components of wind turbines), whereas the modular structure of photovoltaics may favour Chinese manufacturing. Thus, from a developmentalist perspective, concentrating solar would be potentially preferable if it were to remain cost competitive. However, the modular design of photovoltaics enables a similar technology to be used at the scale of solar farms and rooftop solar, whereas concentrating solar favours the larger scale. Thus, there is arguably a valence in the photovoltaic form of a solar-farm design that would be consistent with small-scale, rooftop solar energy development, which then links the photovoltaic farm with the politics of rooftop solar energy discussed above.

At the scale of individual buildings such as rooftop solar photovoltaics, the sociotechnical system can be configured in several ways, of which the following set of ten technical and social distinctions provides a basic map: off-grid systems versus systems that permit grid interconnection, on-site storage versus systems that use the grid as the 'storage' site, systems owned by the building owner or owned by a company that manages the system as part of power-purchase agreement, systems partially funded by a government assistance program or funded by the building owner, and a rate structured by feed-in tariff versus a more market-oriented rule such as avoided cost. It is possible, then, to analyse this set of sociotechnical distinctions with respect to the political field.

Social liberalism. Political leaders who are primarily concerned with redistributive issues and social fairness would be relatively neutral with respect to grid interconnection, storage, and ownership, but they would have a valence toward a government assistance program and a feed-in tariff, both of which would correct for market failures. The first failure is the need to provide a fair source of assistance across income levels, and the second is the need to correct the capacity of the electricity service provider to charge monopsony prices for electricity supplied to the grid.

Socialism. Municipal electricity organizations would tend to favour grid interconnection, on-grid storage, and government assistance programs to provide a fair source of assistance. However, they would likely favour ownership by the public agency rather than the building owner. On the issue of a payment structure, the public organizations would likely prefer a more market-oriented price structure to the inflexibility of a feed-in tariff.

Localism. The central feature of localism, ownership by the building owner, is facilitated by a feed-in tariff and government assistance programs. Local ownership is potentially compatible with grid interconnection and off-site storage, but there is a valence in off-grid systems with on-site storage toward local ownership.

Developmentalism. Because solar photovoltaics in the United States are increasingly imported, the primary form of developmentalism is the creation of jobs in the installation and service industries for solar energy. Thus, leaders who are most concerned with developmentalist issues would support features of a sociotechnical system that favour demand creation. Government assistance programs (but to create demand, not to neutralize income disparities) and a feed-in tariff would strengthen demand. A developmentalist might also support grid interconnection and on-grid storage, because the policies would facilitate demand growth.

Neoliberalism. Because a neoliberal would favour the removal of barriers to the development of markets, a neoliberal would likely support grid interconnection, off-site storage, power-purchase agreements, and a feed-in tariff, but an ideal typical neoliberal would likely oppose government assistance programs. In general a neoliberal would favour systems that enable choices and maximum flexibility in system design so that various types of markets and product strategies could coexist in a competitive relationship.

In summary, the ways in which a specific feature of a sociotechnical system might be viewed as neoliberal require a broad analysis of the distinctions at play in the technological field and their relationship with positions in the political field. There are both possibilities of convergence and divergence in the politics of technology, and those possibilities provide opportunities for political conflicts and coalitions to emerge. For example, an ideal typical neoliberal, social liberal, and developmentalist might favour the need to open up the grid to interconnection from distributed energy sources, but a neoliberal might part ways with both on the issue of tax credits and other incentive programs. In turn the social liberal and developmentalist may both support tax credits, but disagree on their structure (redistributive goals versus demand-creation goals). A neoliberal might oppose the large-scale solar farm because it is associated with a renewable electricity standards and carbon regulation, but net meters may be more closely in line with an approach to renewable energy that enables new market development and frees up barriers to them. In turn, a social liberal could agree that distributed renewable energy is preferable (but due to redistributive goals), whereas a developmentalist might be neutral on the issue or split depending on what kind of demand policy is foremost (supporting local solar manufacturing versus installation).

The politics of sociotechnical system design have implications for the politics of research fields as well. A research funder could choose to favour one or more of a wide range of research programs: concentrating solar versus photovoltaics, off-grid storage systems versus smart grids and grid interconnection, and even research on types of regulatory incentives and ownership programs. From the perspective of a researcher in one of the fields, the relationship between a research program and a political ideology would likely be invisible, because the researcher is preoccupied with the technical questions of the research field. However, when one steps back to an analysis of the relations among distinctions in the technical and scientific fields and the distinctions in the political field, it is possible to see the connections that may otherwise be 'misrecognized', to use Bourdieu's phrase (1981).

To generalize the argument, other than in the social sciences the connection between a political ideology and a technology and associated research field is contingent on specific historical and cultural contexts. This does not mean that the relationship is subject to infinite interpretive flexibility, a view that is suggested by the constructivist and performativity traditions in science and technology studies. Although those approaches provided a helpful corrective to determinist forms of structuralism, the alternative adopted here steers a course between infinite interpretive flexibility and technological determinism. Some choices in the design of sociotechnical systems (and the choices in agendas for associated research fields) have a valence toward a position in the political field. To develop an analysis of the relationship between neoliberalism and the technosciences, one must first begin with the problem of relationships, not only the relationships between neoliberalism and other political ideologies in the political field, but also the relationships among design features in the field of technologies and the priorities for research agendas in the associated research fields. Thus, field sociology offers a methodology for studying the problem of neoliberalism and the technosciences that opens up new research questions and new insights into the politics of technology.

Conclusions

The conceptual and methodological framework outlined here suggests a way of thinking about the history of neoliberalism and its relationship to science and technology that begins with the problem of inter-related, semi-autonomous fields where practices and coalitions are articulated and remade. By starting with a view of neoliberalism as a position in a contested political field, it becomes possible to map the homologies between distinctions in scientific and technological fields with those in the political field. In the case of economics and the policy-oriented social sciences, the homologies are transparent in the schools and research programs. However, in the case of the natural sciences and engineering fields associated with industrial technologies, the politics of technology can only be understand through a double exploration of technological and political distinctions.

In summary, the field sociology of neoliberalism and the technosciences suggests caution with respect to analyses that assume a comprehensive relationship between one type of technoscience and neoliberalism. Although many new research fields and associated technologies have emerged since the ascendency of neoliberal political leaders after the 1980s, the correlation does not necessarily imply a straightforward relationship. A relational and comparative approach makes it possible to see both the political valences toward an ideological position and the potential for neutrality on some design configurations.

The field sociology of neoliberalism also provides new ways to think about its possible future. The fact that there was an historical change during the 1970s and 1980s from the dominance of social liberalism to neoliberalism in many political fields across the world raises the question of what type of ideology might emerge to replace neoliberalism. Historical change tends to create new contradictions and countermovements (Polanyi 1944), but much of the scholarship on a post-neoliberal order is clouded by nostalgia, such as belief that increasing within-nation inequality will lead to a swing of the pendulum back toward the happy days of social liberalism, or by utopianism, such as belief that a crisis in capitalism and global ecology will lead to a great transition to socialism or localism. I suggest that the historical condition of the relative decline of an economic superpower may be leading to resurgence of a modified form of its past of developmentalism. Developmentalist liberalism can be distinguished from neoliberalism and social liberalism on several grounds, of which the increasingly defensive position on trade and the corresponding invigoration of industrial policy are the most central. Developmentalism remains mainstream because it does not challenge the premise of restricted government intervention into an economy based on large, publicly traded corporations.

In contrast with the three mainstream positions, agents in subordinate positions in the political field advance more radical visions of the global economy, such as an economy based on locally owned, independent enterprises or one based on much higher levels of public ownership. Although it is possible that ideologies currently in a subordinate position in the political field may become dominant over the long run, the likely outcome, albeit not necessarily desirable from either a sustainability or social justice perspective, may be a growing tendency toward developmentalism.

The shift within the American political field in the twenty-first century is consistent with its nineteenth-century past and its isolationist proclivities. During the nineteenth century the United States was a secondary economic power with respect to European countries, and it pursued its successful policy of industrialization based on strong tariffs and import substitution. Only after the United States had achieved clear hegemony in the global economy did it shift to a liberalized position on trade, which at that time benefited the country by opening up foreign markets to trade on terms favourable to the United States (Chang 2008). However, as foreign competition increased, sentiment in favour of a more defensive approach to trade increased, first with respect to Japan during the 1980s and then today with respect to China. Thus, the stage is set for the political field in the United States during the twenty-first century to look more like that of the nineteenth century than the twentieth century. Because the United States has been the world's leading supporter of neoliberal policies, a shift toward developmentalism could have global implications for the history of neoliberalism as well as for the global economy.

Acknowledgements

I appreciate comments from Les Levidow and Luigi Pellizzoni on a preliminary draft. The study is based partly on research funded by the Science and Technology Studies Program of the National Science Foundation for the grant titled 'The Greening of Economic Development' (SES-0947429, Grant Number 1156187). Any opinions, findings, conclusions, or recommendations expressed in this book are my own and do not necessarily reflect the views of the National Science Foundation or others who are acknowledged. A few of the quotations used here also appear in Hess (2012).

References

American Public Power Association. 2011. *Public Power: Shining a Light on Public Service*. Available at: http://www.publicpower.org/files/PDFs/PPFactSheet.pdf [accessed: 30 September 2011].

Bourdieu, P. 1981. *The Political Ontology of Martin Heidegger*. Stanford, CT: Stanford University Press.

Bourdieu, P. 1998. *Acts of Resistance: Against the Tyranny of the Market*. New York: New Press.

Bourdieu, P. 2005. *The Social Structures of the Economy*. Malden, MA: Polity.

Chan, S. and Bradsher, K. 2010. U.S. to investigate China's clean-energy aid. *New York Times*, 15 October. Available at: http://www.nytimes.com/2010/10/16/business/16wind.html?ref=united_steelworkers_of_america [accessed: 30 September 2011].

Chang, H. 2008. *Bad Samaritans: The Myth of Free Trade and the Secret History of Capitalism*. New York: Bloomsbury Press.

Chang, H. 2011. The 2008 world financial crisis and the future of world development, in *Aftermath: A New Global Economic Order?*, edited by C. Calhoun and G. Derluguian. New York: New York University Press, 39 –64.

Coley, J. and Hess, D. 2012. *Green Energy Laws and Republican Legislators in the U.S.* Nashville, TN: Sociology Department, Vanderbilt University.

Drajem, M. and Martin, E. 2011. U.S. solar manufacturers request duties on solar imports. *Bloomberg BusinessWeek*. [Online, 20 October]. Available at: http://www.businessweek.com/news/2011-10-20/u-s-solar-manufacturers-request-duties-on-chinese-imports.html [accessed: 17 December 2011].

Eisinger, P. 1986. *The Rise of the Entrepreneurial State: State and Local Economic Development Policy in the United States*. Madison: University of Wisconsin Press.

Fligstein, N. and McAdam, D. 2011. Toward a General Theory of Strategic Action Fields. *Sociological Theory*, 29(1), 1–26.

Foucault, M. 2008. *The Birth of Biopolitics*. London: Palgrave MacMillan.

Harvey, D. 2005. *A Brief History of Neoliberalism*. Oxford: Oxford University Press.

Harwood, J. 2010. 53% in U.S. say free trade hurts country: NBC/WSJ Poll. *CNBC.com*. [Online, 28 September]. Available at: http://www.cnbc.com/id/39407846/53_in_US_Say_Free_Trade_Hurts_Nation_NBC_WSJ_Poll [accessed: 30 September 2011].

Hess, D. 2009. *Localist Movements in a Global Economy*. Cambridge, MA: MIT Press.

Hess, D. 2012. *Good Green Jobs in a Global Economy*. Cambridge, MA: MIT Press.

Hess, D., Banks, D., Darrow, B., Datko, J., Ewalt, J., Gresh, R., Hoffmann, M., Sarkis, A. and Williams, L. 2010. *Building Clean-Energy Industries and Green Jobs: Policy Innovations at the State and Local Government Level*. Troy, NY:

Science and Technology Studies Department, Rensselaer Polytechnic Institute. Available at: www.davidjhess.org/greenjobs.html [accessed: 30 September 2011].

Hindery Jr., L. 2010. China trade reform: Mission (not yet) accomplished. *Huffington Post*. [Online, 1 June]. Available at: http://www.huffingtonpost.com/leo-hindery-jr/china-trade-reform-missio_b_595872.html [accessed: 17 December 2011]

Hughes, T. 1983. *Networks of Power: Electrification in Western Society, 1830–1900*. Baltimore, MD: Johns Hopkins University Press.

Hughes, T. 1987. The evolution of large technological systems, in *The Social Construction of Technological Systems*, edited by W. Bijker, T. Hughes and T. Pinch. Cambridge: MIT Press, 51–82.

Klein, N. 2011. *Keynote Lecture*. Paper to the Annual Conference of the Business Alliance for Local Living Economies (Bellingham, WA). Available at: http://vimeo.com/25839058 [accessed: 30 September 2011].

Krugman, P. 2009. Chinese New Year. *New York Times*, 31 December. Available at://www.nytimes.com/2010/01/01/opinion/01krugman.html [accessed: 30 September 2011].

Labor Research Organization. 2004. *Wages and Benefits: Real Wages (1964–2004)*. [Online]. Available at: http://www.workinglife.org/wiki/Wages+and+Benefits%3A+Real+Wages+(1964-2004) [accessed: 30 September 2011].

Moore, K., Frickel, S., Hess, D. and Kleinman, D. 2011. Science and neoliberal globalization: A political sociological approach. *Theory and Society*, 40(5), 505–32.

Mulkern, A. 2010. Rival ethanol groups campaigning to woo senators, clobber each other. *New York Times*, Online, 13 April. Available at: http://www.nytimes.com/gwire/2010/04/13/13greenwire-rival-ethanol-trade-groups-campaigning-to-woo-s-2028.html?pagewanted=all [accessed: 30 September 2011].

Peck, J. and Tickell, A. 2002. Neoliberalizing space. *Antipode*, 34(3), 380–404.

Polanyi, K. 1944. *The Great Transformation*. Boston: Beacon.

Samuelson, R. 2007. The end of free trade. *Washington Post*, 26 December. Available at http://www.washingtonpost.com/wp-dyn/content/article/2007/12/25/AR2007122500863.html [accessed: 30 September 2011].

Spence, M. and Hlatshwayo, S. 2011. *Jobs and Structure in the Global Economy*. [Online]. Available at: http://www.project-syndicate.org [accessed: 30 September 2011].

Van Nostrand, J. 2011. Legal issues in financing energy efficiency. *Journal of Energy and Environmental Law* Winter, 1–16.

Wald, M. and Savage, C. 2011. Furor over loans to failed solar firm. *New York Times*, 14 September. Available at: http://www.nytimes.com/2011/09/15/us/politics/in-solyndra-loan-guarantees-white-house-intervention-is-questioned.html [accessed: 30 September 2011].

Winner, L. 1986. *The Whale and the Reactor*. Chicago, IL: University of Chicago Press.

Conclusion

Making sense of neoliberalism and technoscience

Marja Ylönen and Luigi Pellizzoni

'In less than one generation neoliberalism has become so widespread and influential, and so deeply intermingled with critically important aspects of life, that it can be difficult to assess its nature and historical importance' (Saad-Filho and Johnston 2005: 1). Perhaps reflecting this difficulty, books, articles and journal special issues devoted to neoliberalism and its environmental, spatial, social and cultural impacts can be roughly divided into two categories: those – the majority – which refrain from drawing any actual conclusion from the essays gathered together, and a minority – the bravest or perhaps the most foolhardy – which venture into such terrain.

Of course, the conclusions drawn also depend on the goals pursued, the approach followed, and the materials collected. Thus, for example, the articles presented in Lave, Mirowski and Randalls (2010) adopt an institutional perspective on the 'neoliberalization of science', dealing with the impacts of public/private partnerships on scientific publication practices, the privatization of stream restoration in the US, and the commercialization of forensic science in Britain and of meteorology in both countries. From these studies the editors draw the conclusions that the character of the university is changing; that the state has been a key player in the promotion of market-based solutions; that science is increasingly produced in response to corporate requirements; that scientific controversies are increasingly adjudicated by the market; and that property rights over, and the commercialization of, knowledge engender elitist patterns and the production of ignorance. Heynen et al. (2007c) deal with a partially different issue – the relationship between neoliberalism and the environment – and include a much broader range of contributions, in number, topic and approach, accounting for an 'incredibly imaginative and frankly disturbing set of experiments, from privatization of wild animals and banking of wetlands to deregulation of water quality' (Heynen et al. 2007b: 287). Their conclusions are consequently set at a higher level of generality, or abstraction. We understand them as follows. First, neoliberalism is a valid analytical category, not least because it works better than capitalism as a means with which to associate academic research and oppositional movements. Second, neoliberalism can be recognized in certain features (such as privatization, creation of markets for 'fictitious commodities', increased role in governance for non-state actors) apparent in a number of cases involving a variety of issues, policies, and environmental changes in different international contexts. Third, despite obvious links between liberalism and neoliberalism, the latter cannot be reduced to the former: 'the question is which strands of liberalism are operationalized, in what ways, and to what effects in different periods' (Heynen et al. 2007b: 290). Fourth, there is no single best way to theorize and study neoliberalism and neoliberalization. Marxian political economy, governmentality, actor-network theory, institutional and ethnographic approaches: these and other outlooks can provide valuable insights. Fifth, despite the variety of neoliberal policies and of their observable outcomes, it is possible to anticipate an overall negative result, 'insofar as their underlying assumptions about markets and property are overwhelmingly mismatched to the character and quality of biophysical systems, and their functions and flows' (Heynen et al. 2007b: 289).

As for the present book, again to be stressed is that it has not sought to furnish any sort of exhaustive overview of the relationship between neoliberalism and technoscience. Its ambitions are conveyed by its title. Aiming to furnish some critical assessments is different from aspiring to give *an* (let alone *the*) assessment. There are many issues and aspects that the book could not address. Yet we believe that it provides a valuable cross-section of the matter. It may be useful, therefore, to return to whence we started and reflect on the insights that the book has generated.

We started by arguing for a close relationship – a tight coupling, or possibly a reciprocally constitutive nexus – between neoliberalism and technoscience. Existing evidence suggested that neoliberalism as a political project and a set of policies, programs, regulations and practices crucially relies on knowledge and innovation-based market competition; and that, simultaneously, technoscience is increasingly shaped by a neoliberal rationality at a variety of levels, from the regulatory and organizational to the motivational and ideational. The task, therefore, was to elaborate on the idea whereby, if neoliberalism transforms technoscience, the latter makes a core contribution to the neoliberalization of society. We also argued concerning the potential insights that arise from different outlooks on technoscience – institutional, political, organizational, cultural and cognitive – and, similarly to Rajan (2006), Heynen et al. (2007a) or Ward and England (2007), we made a case for the legitimacy and usefulness of a plurality of theoretical frameworks, not only across different studies but also within individual analyses. We decided, moreover, to organize our materials around three broad perspectives: the connection among neoliberalism, technoscience and late capitalism and the implications of that connection in two main fields of intervention – humanity and the natural environment. Finally, we remarked that the book was inclined to the opposite side of the micro-empiricist approach widespread in the STS literature. An important choice indeed, as we were aware from the outset. Yet, rather than embarking on methodological discussions or debatable costs/benefits assessments, we decided to take a more relaxed standpoint. Detailed case studies are valuable for avoiding unwarranted abstractions; yet drawing broader implications from this kind of inquiry is no less problematic. On the other hand, a sensible approach to policy fields, technological sectors or regulatory arrangements, or even to entire historical trends, narratives or ideational perspectives, does not prevent, but rather benefits from, looking at concrete examples – as, we believe, the chapters in this book testify.

In short we felt, and still feel at the end of our journey, that there is plenty of room for a variety of approaches. Theoretical and methodological pluralism may lose something in argumentative elegance (even though not necessarily), yet on such an entangled issue as neoliberalism and technoscience it may significantly increase in analytical strength and thought-provoking capacity. Of course, syntheses and distillations become more difficult. Yet we hope not to inflict too great violence on the arguments of the authors, and to be of good service to the readers, if we collect in what follows our general impressions and understandings.

Overall, compared with Lave, Mirowski and Randalls (2010), Heynen et al. (2007c) and other literature intersecting with our topic, the chapters offer – we submit – significant elements of additional reflection. They conduct valuable reviews of the issues addressed and present original findings, ideas and interpretations, increasing the intelligibility of the many facets of the question.

Several pages of the book can be regarded as a reply, from the vantage point of technoscience-focused studies, to the question of what neoliberalism is. We saw that the literature is replete with discussions on its features and its ultimately homogeneous or heterogeneous character. This in part reflects the fact that neoliberalism is not a mode of production including a well-defined set of features like feudalism or capitalism. 'Neoliberalism straddles a wide range of social, political and economic phenomena at different levels of complexity' (Saad-Filho and Johnston 2005: 1). Something on this issue (which is by no means a merely academic one) emerges from the specific outlook of the book. Neoliberalism relates with technoscience at a number of levels. It is an ideology and a narrative informing agents' sense-making in a wide variety of contexts. Yet it is much more than this, for it involves strong institution-building and practice-shaping. To this end it combines different modes or approaches. 'Self-regulatory' governmental mechanisms based on market exchanges and competitive entrepreneurship are flanked and supported by robust injections of the sovereign regulatory and economic power of the state (or its transnational surrogates), first of all in providing legal frameworks for the expansion of property rights. In this sense 'the most basic feature of neoliberalism is the systematic use of state power to impose (financial) market imperatives, in a domestic process that is replicated internationally by "globalisation"' (Saad-Filho and Johnston 2005: 3). The relevance of the rule of law for neoliberal politics has been stressed several times (see e.g. Mattei and Nader 2008). This feature emerges, however, with particular salience in the technoscientific field, with regard to such diverse matters as psychiatric reform, biofuels regulation, carbon markets, green energy development, ICTs and biotech.

Moreover, as a form of government of and through technoscience, neoliberalism exhibits both elements of continuity and change vis-à-vis previous forms of liberalism. This duality significantly contributes to making it an elusive phenomenon. Much depends on the features that can be singled out in different fields, and how the interpreter reads them. On browsing the chapters on biofuels, green energy and psychiatry, for example, but also Reynolds and Szerszynski's account of the dubious import of cutting-edge innovation for the latest thrust of capital accumulation, one is induced to emphasise the elements of continuity. By contrast, on reading the chapters on ICTs, biotechnologies and converging technologies, and even Reynolds and Szerszynski's reflections on the subsumption of science to neoliberal rule, the predominant impression is that of a qualitative difference. In this respect, Blok's chapter finds itself somewhat caught in between the puzzling character of carbon markets vis-à-vis more traditional forms of commodification of nature; an ambiguity that is reflected in the varied positions taken by civil society actors with regard to this policy approach.

The enigmatic or elusive features that neoliberalism exhibits at the interface with technoscience also appear in some emerging tensions as regards the dynamics and outcomes of neoliberalization. The issue here is not so much the more or less consistent and encompassing aims of neoliberal programs and policies, as the concrete picture that results from technoscience-related questions. More than 'varieties' of neoliberalization processes, as often depicted by the literature, we seem to be confronted with major unresolved frictions, compromises or contradictions springing from different subject matters and policy fields. Put otherwise, the frictions, compromises and contradictions on which scholarship builds its argument about neoliberal varieties stand out with particular evidence in this problem area. Many elements of the book convey this impression: Hess's analysis of green energy policy in the US as co-existence of widely differing views of liberal democracy and associated technologies and organizational solutions; Ferreira et al.'s portrayal of *psy* sciences as an internally fragmented field, where psychiatric reforms have an ambiguous relation with neoliberal 'humanism' in their shared appeal to an empowered, autonomized, entrepreneurial person; Arnaldi's account of the convergences and divergences among neoliberalism, posthumanism and Fukuyama's bioconservative liberalism; Gandini's discussion of the role of ICTs as drivers of a culture that originally supported but now moves beyond neoliberalism, opening the way to a new political-economic cycle; Blok's examination of the multiple basis of carbon trading in terms of concerns, aims and grounds for legitimacy; Bard's reflection on the simultaneous emphasis, in neoliberal biopolitics, on human enhancement and human deficiency, individual empowerment and decomposition in body elements or re-aggregation in population dynamics; Levidow, Papaioannou and Birch's account of biofuels policy as a progressive strengthening of the interlinking of market, knowledge and resource exploitation, and simultaneously as the reproduction of the usual dynamics of technoscience policies (with the well-known sequence of problem and solution-framing, citizen and stakeholder contestation, and the reiterated promise of a technical fix); Reynolds and Szerszynski's discussion of the discrepancy between hype and actual mundaneness in accounting for the technologies of post-Fordist accumulation; and finally our own remarks about the ambiguous, intimately contradictory, character of neoliberal rationality, with its combination of a hyper-modernist notion of human agency and a post-modernist account of world pliancy.

However, a tension is also perceptible among the various chapters as regards the emerging or likely outcomes of such ambivalences or contradictions. Are we faced with really diverse and open-ended processes, with possible socially and environmentally 'benign' results; or are varieties and frictions of aims, means and effects to be traced to an underlying logic able to absorb or include any 'overhanging' or 'overflowing' element? Can, for example, the non-market social and normative drivers of carbon trading be subsumed to market mechanisms without losing much of their political and ethical strength? Can developmentalism or current social liberal approaches to the green energy sector be imagined independently of a broader framework, neoliberal in a looser (but no less compelling) sense? Can the forms of sociality spreading within the new ICTs-related intellectual working class remould the capital-dominated task environment in which their capacities are nurtured and exploited? Can the present 'outside' of the asylum represent a really liberating context for the mentally ill, given also that the definition of illness (mental or else) is continuously evolving? Can ethics, public dialogue and science-based social struggles be disentangled from the neoliberal contexts and underpinnings that affect their present declensions to various extents? Can, more in general, the emerging forms of technoscience-connected subjectivation be uncoupled in some really significant respect from neoliberal modes of subjection?

No univocal answers, at least to our understanding, can be drawn from the theoretical and empirical explorations of the book. Yet we consider this to be a significant (if possibly disappointing) result in itself. We do so for two reasons. First, it confirms that the problem of neoliberalism, also with regard to technoscience, is *very* intricate. Second, the stakes involved in a critique of, or resistance against, neoliberalism stand out in all their salience. Can we talk – for example with reference to the governance of biology, ICTs and climate – of reforms, adjustments, or 'more appropriate' uses of existing arrangements in their intertwining of sovereign power, market rule and individual agency, or should we look for major shifts – and who might be the drivers of such shifts, and what kind of shifts should we expect, support or elaborate? And above all, is the actual situation as contradictory and entangled as it seems, or is this impression at least partially conveyed by the inadequacy of our analytical categories?

The intimate relationship between neoliberalism and technoscience, in any case, stands out in a way that seems fully to justify talk of reciprocal constitution. This is particularly so when we consider the distinctive constructivism that emerges at the interface among science, technology and market capitalism. Not only do hype, speculation, promissory anticipations of pending technical fixes or major steps forward in the handling of natural/social problems or threats receive particular prominence throughout the book, but the embroilment of actual capacity with futuristic imagination, and of precise calculation with probabilistic estimation and tendential forecast, intersects with, or takes the shape of, more or less explicit and 'governed' market relationships. This reiterates the question of whether the intensified neoliberal features of modern culture, political liberalism and market capitalism engender some major, qualitative change in social affairs. Technoscience emerges as a possibility condition for the construction of markets, or for the performativity of economics; at the same time, the technoscientific remoulding of space and time, of the biophysical and social world, is shaped around and according to market-dominated or market-oriented cognitive framings and value formations. This two-edged character of current social affairs is conveyed by a number of concepts and arguments presented in the chapters: from the 'homo carbonomicus' of climate change policies to the 'singularities' composing the intellectual multitude of ICTs; from the 'marginal land' of biofuel production – an allegedly amorphous and meaningless material waiting for insertion into the global value chains – to the comparable ways in which individuals, groups and biophysical entities are remoulded in biotechnology research; from the subject and object of desire, will and perfectibility in current posthumanist and liberal (with and without the 'neo' prefix) thinking and policing to the connections among trade, territory, people, industry, public administration and technological innovation in green energy policy; from the relationship between respatialization of production and new enclosures of the scientific commons in the post-Fordist economy to the entrepreneurial citizen of psychiatric reforms or the intertwining of people's individualization and de-individualization within the pervasive frame of 'risk'. These examples show that market and capital are not only the forging but also the forged entities. Technoscience is undoubtedly affected by or shaped according to neoliberal views, policies and regulations, yet it also acts as a powerful agent on its own, both by increasingly giving the market a particular scope or rationale – knowledge and information become the basic resources to be produced and exchanged – and by providing market relationships with shapes, directions and turning points that are not produced by these same relationships; by giving, in other words, form and salience to a variety of non-human 'actants' that retroact on human goals and expectations. Consider, for example, the partially 'governed' but partially 'unmanageable' development of information networks or post-genomic identities; the technoscientific underpinnings of the economic, political and cultural elaborations of carbon trading; the technology-driven imaginations or spaces of possibility that operate as (prospective) 'hard facts' in posthumanist discourses and policies.

What is at stake in the close coupling or mutual constitution of neoliberalism and technoscience is, by any evidence, not only the control over, but the intimate constitution of, humanity and nature. In this sense, too, the intensification of l ong-established features of modern culture and market capitalism raises a problem of interpretation. On the one hand, the dominant discourse of natural and social threats, and the diffused dread that it engenders through a combination of economic instability and headlong innovation, leads to a technoscience-mediated and technoscience-forging reorganization of social relationship under the heading of 'security' (at any level: interpersonal, corporate, urban, national, global). On the other, it is the very constitution of what humanity and nature ultimately are that is undergoing major shocks. What is striking, therefore, is not only that nature or science are 'neoliberalized', or that neoliberal ideas and programs are shaped according to technoscientific possibilities and imaginaries, but that the traditional distinction between right and left, conservative and progressive orientations and arguments vacillates on these questions. Indications of the vanishing of traditional lines of division frequently emerge throughout the book, and with regard to such diverse issues as, for example, psychiatry and carbon markets. Of course, this vanishing does not at all correspond to the widespread rhetoric of the disappearance of any (reasonable) divide within a unified worldview or a joint effort towards common goals – fighting terrorism, the economic crisis, the environmental threats (on this rhetoric see e.g. Mouffe 2005, Swyngedouw 2010, Žižek 1999). On the contrary, conflicts around the world erupt along different lines of friction. If we consider, for instance, the struggles over the use of land and nature – with topics ranging from infrastructure and farming practices to the free access to water and genetic diversity – we see that adversary positions are hardly amenable to traditional progressive and conservative orientations. The defence of resources, places and cultures against their appropriation or inclusion in globalized capital dynamics finds radical left and radical green groups often allied with conservative populist, religious and localist formations against cosmopolitan liberals, 'third way' social democrats, 'old' industrial leftists, and technocratic greens. This reshaping of the political spectrum in current contentious politics requires major interpretive efforts – starting, for example, from acknowledgment that the shift from Fordism to post-Fordism and from the welfare to the workfare state produces novel forms of alliance, not according to labour vs. capital or regulated vs. free market aggregations, but according to issues of risk and opportunity, openness to and protection against the challenges of global markets and financial capitalism (see e.g. Azmanova 2010). Yet if, at the most basic level, what is at stake in many conflicts is the meaning of and the means for progress, development, freedom and emancipation, then technoscience appears of paramount importance. It is therefore against the specific aims and forms of science and technology that the present meaning and social constitution of progressivism and conservatism are to be gauged.

In sum, there is much work awaiting social science scholars. It is usual for a book to conclude by expressing a call for further inquiry. In our case such a call is anything but rhetorical. This collection of essays has offered insights into a truly complex and major issue. It is possible that the time of neoliberalism is coming to an end, or that this fragmented yet encompassing social project is undergoing another major evolution. The world 'after neoliberalism', however, is most likely to be one 'in which neoliberalization, in all variants, all its guises, all its hybrid formations, continues to cast a long shadow over matters of social and economic justice' (England and Ward 2007: 260). And over matters of technoscience, we may add; but bearing in mind, once again, that the opposite also applies: technoscience is bound crucially to affect any further evolution and transformation of, or exit from, the neoliberal world.

References

Azmanova, A. 2010. Capitalism reorganized: Social justice after neo-liberalism. *Constellations*, 17(3), 390–406.

England, K. and Ward, K. 2007. Conclusion: Reflections on neoliberalization, in *Neoliberalization: States, Networks, Peoples*, edited by K. England and K. Ward. Oxford: Blackwell, 248–62.

Heynen, N., McCarthy, J., Prudham, S. and Robbins, P. 2007a. Introduction: False promises, in *Neoliberal Environments: False Promises and Unnatural Consequences*, edited by N. Heynen, J. McCarthy, S. Prudham and P. Robbins. London: Routledge, 1–21.

Heynen, N., McCarthy, J., Prudham, S. and Robbins, P. 2007b. Conclusion. Unnatural consequences, in *Neoliberal Environments: False Promises and Unnatural Consequences*, edited by N. Heynen, J. McCarthy, S. Prudham and P. Robbins. London: Routledge, 287–91.

Heynen, N., McCarthy, J., Prudham, S. and Robbins, P. 2007c. (Eds.) *Neoliberal Environments: False Promises and Unnatural Consequences*. London: Routledge.

Lave, R., Mirowski, P. and Randalls, S. 2010. (Eds.) STS and Neoliberal Science. *Social Studies of Science*, 40(5), 659–791.

Mattei, U. and Nader, L. 2008. *Plunder: When the Rule of Law is Illegal*. Oxford: Blackwell.

Mouffe, C. 2005. *On the Political*. London: Routledge.

Rajan, K.S. 2006. *Biocapital. The Constitution of Postgenomic Life*. Durham, NC: Duke University Press.

Saad-Filho A. and Johnston, D. 2005. Introduction, in *Neoliberalism. A Critical Reader*, edited by A. Saad-Filho and D. Johnston. London: Pluto, 1–6.

Swyngedouw, E., 2010. Apocalypse forever? Post-political populism and the spectre of climate change. *Theory, Culture & Society* 27(2–3), 213–32.

Ward, K. and England, K. 2007. Introduction: Reading neoliberalization, in *Neoliberalization: States, Networks, Peoples*, edited by K. England and K. Ward. Oxford: Blackwell, 1–22.

Žižek, S. 1999. *The Ticklish Subject. The Absent Centre of Political Ontology*. London: Verso.

Index